Water Quality Modeling

Mervin D. Palmer

A GUIDE TO EFFECTIVE PRACTICE

The World Bank
Washington, D.C.

Manufactured in the United States of America
First printing May 2001
1 2 3 4 05 04 03 02 01

Cover design by Tomoko Hirata

Library of Congress Cataloging-in-Publication Data

Palmer, Mervin D., 1937-

Water quality modeling : a guide to effective practice / by Mervin D. Palmer.

p. cm.

Includes bibliographical references.

ISBN 0-8213-4863-9

1. Water Quality—Mathematical models. 2. Water quality management—Mathematical models. 3. Water quality—East Asia—Mathematical models—Case studies. I. Title.

TD370 .P35 2001

363.739'42'015118—dc21 00-049951

Contents

Tables

Figures

Acknowledgments

Preparation of this guide was financed by the East Asia Social and Environment Sector Unit (EASES) under the direction of Zafer Ecevit. The guide is the product of an extensive review carried out by Merv Palmer under the guidance of Glenn Morgan and Jack Fritz. The report has benefited greatly from the contributions of others with diverse perspectives on water quality prediction and management. In addition, the preparation of the report has drawn extensively on the experience of institutions worldwide active in the development of water quality prediction models.

EASES gratefully acknowledges the significant advice provided by the following individuals: Geoffrey Read, Edouard Motte, and Wiebe Moes of the East Asia Urban Development Unit. Doug Olson of the East Asia Rural Development and Natural Resources Unit provided technical inputs on case studies and contributed many hours of peer review. Heinz Unger, Rob Crooks, Anil Somani, and Patchamathu Illangovan from EASES provided valuable insights during the conceptualization of the report. Editorial assistance in design, layout, and preparation of illustrations was provided by Mellen Candage, Catherine Fadel, Kaye Henry, and Nicola Marrian.

Foreword

The quality of surface water resources affects virtually all aspects of life in East Asia. Pollution from industrial, agricultural, and domestic sources continues to grow throughout the region, affecting rivers, lakes, estuaries, and coastal regions. As population growth with its attendant economic growth continues throughout the region, water resource managers must increasingly use analytical tools to assist in the formulation of sustainable water management strategies. One family of analytical tools that can be of tremendous value are water quality prediction models.

Experience demonstrates that the design, implementation, and monitoring of waste management schemes can benefit from the use of numerical modeling. The evolution of these prediction tools now creates many more opportunities for operational and economic optimization than were available a generation ago. Computer systems and software are less expensive, more accessible, and easier to use than ever before. At the same time, increased accessibility creates more opportunity for misapplication under field conditions. To be most effective, water quality prediction models must be used in ways appropriate to the task at hand; they require the expertise of knowledgeable technical specialists and reliable input data. Water quality prediction can be expensive and potentially inconclusive if not approached in a systematic manner.

This technical guide provides a review of the state of water quality prediction models available to the practitioner today. It provides essential technical background for individuals who may be required to develop models or interpret their results in the context of project design. The guide, which is designed as a more in-depth and revised treatment of this issue as presented in the World Bank's *Pollution Prevention and Abatement Handbook 1998,* will fill a significant gap in the technical literature. More important, it demonstrates how these models have been applied to recent World Bank projects.

It is hoped that this publication will provide a technical discussion of water quality prediction that is accessible to practitioners in developing countries. We believe this will become an essential guide for anyone involved in the design or use of water quality prediction models. In this way, the guide will contribute to the more effective use of prediction tools and will improve the quality of water projects in general.

Zafer Ecevit
Sector Director
Environment and Social Development Unit
East Asia and Pacific Region
The World Bank

Executive Summary

The challenge of understanding and managing surface water quality problems in East Asia is growing. Those responsible for managing water resources must look to a variety of analytical tools to design, implement, and monitor sustainable water quality management programs. An important family of tools are numerical water quality prediction models. With the availability of powerful desktop computers and the growing availability of software, these tools have become increasingly accessible.

Numerical models have demonstrated an impressive capacity to support important water resource decisions in developed countries. Models are typically used to support development and public policy decisions in a variety of areas: simulation of discharges, outfalls, and intakes; changes to wastewater treatment systems; approval of changes in industrial processes; operation of dams and reservoirs; and water resource allocations, among other uses. The value of modeling is important in economic and financial terms with regard to determining particular project options and phased investment programs.

Not all models, however, are appropriate under all conditions. They vary greatly with respect to their analytical approach, underlying assumptions, data needs, and output capacity. In the context of developing countries, the utility of models must be carefully examined in the light of important constraints such as lack of experienced technical staff, poor-quality data sets, and lack of or poorly enforced quality control protocols.

This document serves as a guide to the utility and relevance of water quality prediction modeling. It draws upon examples from recent World Bank water resources and wastewater management projects. The goal of the guide is to provide a broad-based understanding of the water quality prediction process and to evaluate the relative merits and cost-effectiveness of using water quality models under field conditions. The guide builds on and revises the chapter on water quality modeling prepared for the World Bank's *Pollution Prevention and Abatement Handbook, 1998.*

The guide does not address groundwater or air quality models. The characteristics of such models are similar to those of water quality models; consequently, understanding water quality models will make it easier for users to become knowledgeable about groundwater and air quality models. For more information on groundwater models, readers may refer to the World Bank publication, "Groundwater: Legal and Policy Perspectives"(technical Paper WTP 456, 1999). For information on air pollution, readers may refer to the Bank publication, *Urban Air Quality Management Strategy in Asia: Guidebook 1997.*

The guide is designed for a range of practitioners, including Bank task managers, environmental specialists, and counterpart technical staff involved in the design, evaluation, and monitoring of sustainable water resources programs. It is not intended to be a comprehensive review of all available water quality prediction models, nor to be an endorsement of specific models. While some typical examples of models are discussed in the report, it provides a thorough review of the current approaches to water quality modeling and the types of parameters that are typically modeled, and offers an assessment of the state of water quality modeling for each parameter.

To illustrate the myriad ways in which water quality prediction models have been used in practice, the guide presents a number of case studies from recent development project with World Bank financial support.. Each case study describes the manner in which models have been used, the constraints encountered under field conditions, the results obtained, and the cost-effectiveness of each application. While the case studies are drawn from East Asian experience, they are relevant to and instructive for all regions.

" — quite simple mathematical equations can model complex systems — "
" — sensitivity depends on initial conditions — "

James Gleick, 1987 in "Chaos"

The guide fills an important gap in the available literature on water quality prediction models. It is designed to cover technical material normally found in more exhaustive, but less accessible, textbooks. In addition, it attempts to bring together information on models that are currently available only through an assortment of difficult-to-find technical references. Potential users of water quality models will find in this guide the basics of commonly used models and will learn how they can be effectively applied in a project context. The guide addresses several inter-related questions:

- What types of models are available and how are they structured?
- What are the basic parameters that models can predict, and how effectively can these parameters be modeled?
- What specific models are available?
- Under what circumstances will water quality prediction models be most beneficial?
- What are the cost and other practical implications of using models in a project setting?

The guide comprises five sections. Chapter 1 provides a general overview of the use of water quality models, including the objectives of water quality modeling, the approach to water quality prediction, the costs of modeling processes, and the general components of typical water quality models. Chapter 2 discusses the most common water quality parameters that are modeled, the receiving water processes, quality assurance and control for the water quality data and model predictions, and the required model resources. Chapter 3 describes generic components of water quality models. In this description, equations are presented. It is not necessary that the reader understand these equations fully, or their method of solution, but it is important to understand the complexity of the model predictions and the requirements for site-specific data. Some prediction models are then discussed, with detailed summaries of these models presented in an Appendix.

Chapter 4 summarizes the present uses of water quality models and provides summaries of some recent Bank development projects that used water quality models. These project summaries include the costs of the modeling processes and, wherever possible, project cost savings resulting from the use of modeling. Chapter 5 discusses the model data requirements and prediction issues such as limited site-specific data, challenges of non-point sources, designs for a water quality monitoring program to support the model, and spill modeling.

Water quality model predictions can be robust and beneficial under a wide range of circumstances. The state of the art is sufficiently advanced for models to be applied in a range of physical settings and for a range of parameters. Model application can be cost-effective, seldom exceeding a few hundred thousand dollars, even for the largest projects. Costs seldom exceed 1 percent of the capital costs in a new water resources project; they have typically cost less than 0.1 percent of facility costs for water management projects. For the seven development projects discussed in this guide, average modeling costs were 0.25 percent of the Bank project funding, including the costs of collecting the site-specific data (which generally accounts for 50-70 percent of modeling costs). The currently available models are generally applicable in developing country situations but can be limited in their effectiveness, primarily as a result of prediction accuracy problems and lack of data.

The guide concludes that models are most effective when certain preconditions are met, specifically:

- The objectives of modeling and their prediction are clearly specified and models appropriate to those objectives are applied. It is preferable that objectives be defined through a stakeholder analysis process.
- Model applications are implemented in a phased, incremental manner using such techniques as simplifications and sensitivity analysis. Staging a project dramatically improves the effectiveness of modeling.

" — in a complex system, when you have parts you no longer have the whole. And the whole cannot be reconstituted from the parts. — there is a move towards more holistic ways of looking at things. Does meteorology depend on a series of pinpoint measurements or an overall view of patterns and processes?"

Edward de Bono, 1990 in "I Am Right You Are Wrong"

- Experienced staff trained in the application and use of models are available.
- Baseline data collected with known quality control protocols are available.
- Precision and accuracy of water quality predictions are quantified using quality control procedures that are appropriate for the objectives.

CHAPTER 1

General Overview of Water Quality Modeling

In most surface water resources projects, there is a need to predict receiving water quality. Some of these projects will be discussed subsequently to show the variety of projects requiring water quality predictions. **Instruments for predicting or simulating water quality are called water quality models.** These models predict or simulate receiving water quality resulting from effluent/contaminant discharges or releases and/or non-point sources for various types of receiving waters' (rivers, oceans, lakes, etc.) characteristics and meteorological conditions. **In most receiving waters, the water levels, currents, flows, temperature, and quality vary with time and location. Similarly, the biochemical processes that affect receiving water quality also vary with time and location.**

All domestic wastewater discharges have large temporal variations (typically 400-1,000 percent during the day) (Metcalf & Eddy, 1991). By using models, it is possible to integrate all the temporal and spatial variables into the water quality prediction. For receiving waters with many discharges at different locations, a computer model must be used.

Because of the variability of modeling parameters, a generalized universal model must be complex. Complex models require highly trained technical staff; consequently, simpler models that consider only the most important processes have become more popular. Some models have been developed for a special situation and others as simplifications of other models. Today, the number of models available to predict water quality is large and growing. In addition to predicting receiving water quality, the models have also been found to be useful diagnostic instruments for water quality management.

This guide discusses a selection of water quality models and documents a variety of applications of models in projects funded by the World Bank in East Asia. The guide does not endorse certain models, but rather examines the important features of different models and shows how some have been used successfully in East Asia Bank projects.

In general, models predict the transport and dispersion processes first, then feed the results into a water quality component of the model. All the processes are represented by equations, and these equations have coefficients or other parameters. Complex models have many components with many coefficients. At the outset, one must have an overall understanding of the modeling process. To this end, schematic diagrams are useful and, luckily, are included in many of the published manuals on water quality models.

Clear objectives of the modeling process must be defined. The World Bank typically uses models to establish priorities for reduction of existing wastewater discharges or to predict the effects of a proposed new discharge. *The World Bank Environmental Assessment Sourcebook* (World Bank, 1991) specifies that the effluent limitations on loadings for each water quality parameter be determined by mathematical modeling. Mathematical modeling techniques have proved to be powerful water resource management procedures. The preface to one of the model manuals (WASP4, 1987) identifies the generic use of a model. "As a diagnostic tool, it permits the abstraction of a highly complex real world. Realizing that no one can ever detail all the physical phenomena that comprise our natural world, the modeler attempts to identify and include only the phenomena, be they natural or man-made, that are relevant to the water quality problem under consideration. As a predictive tool, mathematical modeling permits the forecasting and evaluation of the effects of changes in the surrounding environment on water quality." Both the so-called point and non-point loadings can be included in the modeling process.

Obviously, both the project objectives and modeling objectives must be clearly defined before any model can be selected and applied. Below are some typical objectives.

- to achieve a certain water quality concentration at a particular location for all time;

- to identify the most cost-effective method for enhancing receiving water quality;
- to reduce human health risk; and
- to reduce the abundance of aquatic plants.

Defining the objectives of the modeling process is important and should involve discussions with stakeholders, regulating agencies, and technical personnel. Most regulating agencies have established water quality objective concentrations for various water quality parameters as well as effluent discharge regulations. The goal is always to achieve the receiving water quality objectives and improve upon them if possible. In simple cases where one point discharge dominates the receiving water quality, it is possible to determine how these receiving water objectives can be achieved and to use a simple, one-dimensional water quality model. In more complex cases, it is much more difficult to determine which water resource management strategy should be used and even more difficult to predict receiving water quality.

During the modeling process, modelers and personnel responsible for the data collection must cooperate and work together with mutual understanding. Because the complexity of cases varies, one effective approach is to achieve the ultimate objective in stages, with each stage requiring its own objective. The objective of stage one, for example, might be to determine the relative magnitude of all upstream loadings on water quality degradation or to reduce the densities of indicator bacteria in an area by 60 percent. Staging a project dramatically improves the effectiveness of the modeling process and its precision and accuracy. It also permits the model user to understand progressively the receiving water quality environment and the effectiveness of the models in simulating the quality in the receiving water. **For complex situations, always divide the project into progressive stages. Carefully define the project and model objectives for each stage.**

In general, models are calibrated using data collected at a particular site and verified against another similar data set. But not all water quality model applications require detailed site-specific water quality monitoring data; examples are applications aimed at developing a better understanding of a complex water resource system, developing a suitable water quality monitoring system, and comparing the effects of different management alternatives, such as land development and land use, on water resources.

Water resources consist of surface water and groundwater. The two are interdependent; surface water is the source of water supply for groundwater, and in many instances groundwater is a source of water for the surface waters. This guide focuses on surface water modeling.

MODELING COSTS

The economic implications of the application of water quality models can be significant. In the early 1980s, the United States spent US$50 million annually on water-related mathematical models used in planning billions of dollars worth of annual water resources investments and managing hundreds of billions of dollars worth of existing facilities (Wurbs, 1995). **Modeling costs constituted about 1 percent of the capital costs for new water resources projects and less than 0.01 percent of facility costs for such projects**. For the seven Bank East Asia water quality model projects summarized in Chapter 4, **the average modeling cost was 0.25 percent, including the costs of data collection**. In many projects, it is difficult to determine costs for water resources sustainability, water use interference, land development, and land use. New water resources projects must be developed in a manner that sustains the water resource and does not degrade it for other future water uses. This is particularly important in the countries of East Asia, which have large population densities and limited water resources. Different project design options must be evaluated using models to determine which will best sustain the water resource.

In many parts of the world, it is a legal requirement to demonstrate that a proposed water resources project will not adversely affect existing water users. This process is referred to as "due diligence." Whether a model is being used to determine wastewater treatment requirements, to locate and design wastewater outfalls, to manage spill abatement procedures, or to assess or evaluate the effects of different land development and land use options, the proponent of such a project must exercise due diligence and be able to demonstrate that the proposed model is technically appropriate. To determine the wastewater treatment requirements, models are required. Models are also used to locate and design wastewater outfalls. For the management of spill abatement procedures and assessment of environmental impact, models are used. And models must be used to evaluate the impact of different land development and land use options.

GENERAL WATER QUALITY MODEL COMPONENTS

Most of the water quality prediction models in use today have the following components:

1. movement in the receiving water;
2. movement, dilution, and dispersion of dissolved substances;
3. first-order decay of dissolved substances;
4. water quality processes; and
5. sediment transport.

None of the components is independent. Component 2 requires the output of Component 1. Similarly, Components 3, 4, and 5 require the outputs from Components 1 and 2. In the model, equations represent the processes within each component. These equations can be time-varying partial differential equations in one-, two-, or three-dimensional space, which is the most complex mode, or other types of equations or segment (box) continuity model systems. A water quality prediction is achieved by solving the appropriate equation(s).

Water quality models tend to be complex, primarily because the equations for water movements are complex. Water movement characteristics are normally determined by numerically solving the partial differential equations of motion and continuity on a grid or element or segment basis. To obtain a mathematical solution, it is necessary to define boundary conditions, and if the model is dynamic (time-varying), initial conditions must also be defined. The physical size of the grids or elements and the time step for the solution must be selected to ensure that the mathematical solutions are stable and converge rapidly. In this instance, the grid size is a function of the mathematical method used to solve the equations and not the characteristics of the receiving water. Normally, there are many more grid points and elements than necessary for a receiving water quality model even when there are many discharges and non-point sources. Because Components 2 to 5 require the outputs from Component 1, the solutions to the equations are carried out on the same grid or elements as used in Component 1. There are techniques for solving the equations in Components 3 to 5 which can use every second, third, etc., grid point; nevertheless, the number of solutions for the water quality predictions are at least of the same order of magnitude as the water movement predictions. The mass balance equation requires averaging over time for a numerical solution. In other words, it is assumed that the concentration of the water quality parameter is constant for the time step in the solution. The validity of this assumption should be checked.

The implications of this water quality model structure affect the applications of the model in different ways as described below.

- For the modeling process to be successful, personnel responsible for the models and for the data collection must function in a cooperative and supportive manner with the confidence and support of the client.
- The water quality data sets required for calibration and verification of the water quality model are large and comparable to the data sets required for the water movement component. These water quality data sets are seldom available and require expensive specialized monitoring programs. For example, photosynthesis and respiration require recording dissolved oxygen and temperatures for 30 hours at several different locations during different times of the year.
- The number of coefficients required for the water quality equations is large. Many of the coefficients are difficult to measure in the field, and some, like the bottom roughness coefficient for flow, resuspension of bottom sediment, and pore water dispersion, are nearly impossible to measure.
- Because the number of coefficients required and the number of data for the boundary conditions are so large, calibration and verification of the water quality model are tedious trial-and-error procedures. In most

instances, the calibration process is only carried out to the level of producing reasonable results for a few data sets (Wurbs, 1995). The predictions from most water quality models should not be considered absolute. Absolute values of any prediction model can be trusted only after extensive calibration and verification (SWMM, 1988).

- If a water quality model can be developed without using a complex water movement model, the model will be easier to calibrate, verify, and apply. It is important to select the simplest model that will satisfy the prediction requirements (World Bank, 1998; SWMM, 1988). The project objectives should be clearly defined before any model is selected. If it is necessary to develop a water management plan for a large number of discharges, the model must predict the receiving water quality for these discharges.

TYPICAL WATER QUALITY MODEL APPLICATIONS

Water quality models (both spatially and temporally variable) are used extensively in Europe and North America for the following:

- *In the approval process for a new discharge outfall or intake, for changes to a wastewater treatment system, for changes in mass loadings, for changes in plant processes, and for ocean disposal.* In each case, the proponent must demonstrate that the receiving water quality is not degraded and other existing water uses will not be adversely affected. This demonstration requires the use of site-specific data and prediction models for a variety of different conditions (e.g., 20-year low flow, spring tide, summer stratification).

- *In land development and land use.* The project proponent must show what water quality changes will occur as a result of the proposed development and what effect the land development or land use will have on the existing water uses. The project proponent must use site-specific data and predict the water quality changes using models.

- *In the approval process for dams and in their operation.* The project proponents are required to demonstrate that the construction of the dam will not adversely affect the water quality either upstream or downstream from the dam site. Operating procedures for the dam are an integral part of the approval process; consequently, the operator of the dam must show that the operating procedures will not adversely affect the downstream and upstream water uses. Extensive site-specific data are required for the approval process, and the proponent must use some of the most sophisticated prediction models available today. In addition to the water quality aspects, the proponents must demonstrate that the safety aspects of the dam have been properly addressed. Dam construction and operation are extensively regulated.

- *To resolve water use conflicts like degraded water quality at a water intake, diminished commercial fish stocks, degraded water quality for other water uses, nuisance growth of aquatic plants, and contaminated well water.* Water quality models are routinely used to resolve such conflicts, using extensive site-specific data both historical and recent. In many instances, the conflict resolution is decided in the courts. The modeling process must be thorough, rigorous, and transparent. Generally, only water quality models available in the public domain can be used, because the models must be made available to all parties in the dispute.

- *In the allocation of water resources to different water uses like drinking water, process water, irrigation, fisheries, and recreational facilities.* Allocations of water resources must be determined through modeling because the allocations are based on a design criterion such as the 50-year low flow, which cannot be measured. To improve the credibility of the model predictions, use only well-tested models. The hydrology in the model must be technically strong. Large data sets are normally required for these models. Water allocation projects are generally undertaken with the same rigor and thoroughness as with dam projects.

- *In the operation of irrigation withdrawals.* Historically, water withdrawals for irrigation have been established based on water availability; however, with the demand for water resources, these withdrawals are being reviewed because of water quality concerns. The prime concern is that the downstream in-flow stream needs to support a viable fishery. Water quality models are extensively used to predict downstream water temperatures and water quality. These models require extensive site-specific water quality and stream characteristics.

- *In spill management.* Models are used primarily for coastal spills of oil to assist in the allocation of remedial measures. These models are not very sophisticated and use very little site-specific data. Prediction models have also been developed for rivers; they make it possible to warn the owners of intakes downstream of the arrival time of a spill. These models are also crude and use little site-specific data.

Some recent examples of the use of water quality models in World Bank projects are summarized in Chapter 4.

CHAPTER 2

Water Quality Model Structure and Processes

BASIC DEFINITIONS

In order to move to a modeling context, it is necessary to agree on a number of basic definitions. This first set of definitions is very broad in that they set the physical context for the analysis.

Spatial characteristics. All models have spatial properties in a fixed grid system, with up to three perpendicular axes, called a Eulerian system, or a moving frame system, called a Lagrangian system, as follows:

One dimension – typically distance downstream or upstream in a river or downstream in an effluent plume;

Two dimensions – typically x and y coordinates in a shallow lake or wide river, or x and z (depth) coordinates in a narrow deep river, lake, or estuary; and

Three dimensions – typically x, y, and z coordinates in large rivers, lakes, and oceans.

Simple models are easier to calibrate, verify, and use, and require less site-specific data. The one-dimensional model is the preferred model. Three-dimensional models, although intrinsically appealing, should be avoided whenever possible because of the large quantities of site-specific data required to ensure the reliability of the model predictions. It is important to keep the model as simple as possible, but the model must fulfill the prediction requirements (World Bank, 1998, p. 119). Sensitivity analysis combined with historical site-specific data can be very useful in simplifying the model.

Lagrangian systems move with a segment of receiving water. A simple analogy helps to differentiate between Eulerian and Lagrangian systems, namely, the speed characteristics of a moving train. In the Eulerian system, details of the train movement would be obtained from observations at fixed points on the track at various times. In the Lagrangian system, details on the train movement would be obtained from observations by someone on the train. The data collected by the two systems are different, and while there are mathematical methods for approximately converting one type of data to the other type, a model is normally developed in the system that is most suitable for the phenomena being studied and applied in that system. Typically, Lagrangian systems are used for spill, tracer, or plume models, and all other applications use Eulerian systems. The Eulerian dynamic river models use time of travel, which is a Lagrangian measurement, rather than velocity. In this case, it is an Eulerian system using time scales from Lagrangian studies. Lagrangian models are generally not suitable for time-varying discharges. The most used Lagrangian models are two-dimensional. It is very easy to use Lagrangian models as stochastic models simply by successively releasing a large number of water particles at a fixed point, then tracing the particles and analyzing the particle statistics at locations in the modeled area.

Temporal characteristics. Models can be steady-state, where time does not appear in any of the model equations, or time-variable, with time as a variable in the model equations. Including time in the equations makes the model much more complex and increases the need for site-specific data for calibration and verification. All time-varying models can be used as steady-state models. In many applications, it is not necessary to use time-varying models. For example, a steady-state model could be used for conditions averaged over a 24-hour period if photosynthesis and respiration are factors, or for tidal averaged conditions. Several different steady-state conditions could also be run for different conditions, and the predictions from each run averaged. For example, a steady-state model could be used to assess the effect of a domestic wastewater discharge by predicting the receiving water quality for the peak discharges at morning, mid-day, evening, and midnight; then the model predictions for these four runs could be averaged for the daily mean. A steady-state model could also be used to predict the receiving water quality for high and low tide slacks, with the rising and falling tides then averaged for a tidal average prediction. **Multiple runs of steady-state models to represent variations in time are simpler than time-varying models.**

If time is a variable in the model equations, it is necessary to define the time intervals of interest. In most instances, the shortest time period is defined by the requirement to obtain solutions to the model equations so that they rapidly converge, and the longest period is defined by receiving water characteristics. If the model is to predict the effects of photosynthesis and respiration, it must predict for a 24-hour period. Similarly, if tidal effects are important, the model must predict for at least two tidal cycles. If low flow is a factor, the model must predict for the period of the low flow condition. If eutrophication is a factor, seasonal model predictions may be required for nutrients. Furthermore, the water quality parameter being modeled may influence the selection of a time period for the model. Some receiving water quality objectives are specified as instantaneous minimums, like dissolved oxygen (DO), unionized ammonia, copper, etc., and some objectives are specified as averages, like biological oxygen demand (BOD), indicator bacterial densities, etc. The water quality objective defines the model prediction requirements. **The time period for the modeling must be compatible with the water quality objective for the water quality parameter being modeled, e.g., single value or an average.**

Non-point sources. Non-point sources are areal discharges to the receiving water such as surface runoff, groundwater discharges, or atmospheric loadings. **Non-point source loadings are frequently important in receiving water quality management in both urban and rural areas and can be easily incorporated in the prediction model** as a point discharge to the model element or river reach. Computationally, treating non-point sources like this is not a problem in water quality modeling. The biggest problem in non-point sources is determining the magnitude of the non-point source loading, because **non-point sources are difficult to measure**. Some detailed hydrological models can predict the surface and subsurface runoff reasonably well and provide various options for ascribing concentrations to runoff and quantifying the non-point loadings. The model documents do provide literature references to assist in the selection of the approach to be used in the model. Ideally, some site-specific data will assist the model user in selecting the non-point source loading method that is the most suitable for a particular application.

To estimate the magnitude of the non-point source loadings in a particular application, some simple mass balances can be computed for the receiving water using only the point discharges and receiving water quality and flow measurements. If these mass balances differ significantly, these differences can be assumed to be non-point source loadings. Another method to quantify the magnitude of the non-point source loadings is to carry out water quality surveys during dry and runoff periods. Historically, it has been found that non-point source loadings of indicator bacteria, nutrients, and metals are frequently important in receiving water quality management for rivers, lakes, and reservoirs. If non-point source loadings are important factors in the water quality model, the model predictions will be less precise, owing to uncertainties associated with quantifying non-point source loadings. In these cases, a less complex model would probably be more suitable.

Water quality monitoring requirements. Water quality monitoring is an integral component of the modeling process (World Bank, 1998). To calibrate and verify the model predictions, site-specific data are required. All the models have coefficients and rate constants that customize the model for a particular application. These coefficients and rate constants are defined in the calibration process using site-specific data. More complex models have more coefficients and rate constants; consequently, simpler models require less site-specific data. It is possible to use many of the models with minimal or no site-specific data by selecting values for the coefficients from ranges of values provided in the model manuals. Nevertheless, **the most precise and accurate model prediction is achieved when there are site-specific data sets for calibration and verifications.**

If different data sets are available for the calibration and verifications, the predictions will have the best precision and accuracy available with that particular model. In practice, this situation is achieved only in large projects that have the appropriate financial and technical resources available. In many instances, the available data set is incomplete, and the remainder of the model data requirements must be obtained from the manual. For these applications, it is important that the model user determine the sensitivity of the model predictions to using the range of values in the manual. One simple method is to use the model with the center third of the range of values in the manual and compare the predictions for the high and low values. **If literature values are used for the model coefficients, the sensitivity of the predictions using those coefficients must be quantified.** If no site-specific data are available, only simple prediction models should be used and a sensitivity analysis should be carried out on the range of coefficients used even if two different model predictions are made and differenced. In other words, even if the model is being applied in a comparative manner, it is necessary to know if a difference of, say, 10 percent is greater than the sensitivity of the model prediction. Using pair differencing increases the precision of the predictions,and the sensitivity determination should follow standard statistical procedures for matched pairs testing.

Models are very useful for designing a more efficient and relevant water quality monitoring system. The model can be used to determine where, when, and what water quality parameters should be measured. Furthermore, the model can be used to identify which discharges should be monitored and what other parameters should be measured to improve the model predictions.

Computational grid. Most of the water quality models consist of partial differential equations that are solved numerically. The grid or element size and time step are selected so that the solution is stable and converges rapidly. Numerical water quality predictions are available at the nodes of the grid. There is some numerical dispersion or imprecision introduced by using a grid, but this imprecision is normally small compared to the imprecision of most of the other water quality parameters in the model.

REQUIRED RESOURCES

Most water quality models can be run on personal computers with at least 64K RAM, 2Meg hard-drive, mathematical co-processor, 3.5" disk drive, and printer. With the rapid growth in computer hardware, most personal computers have more than enough capacity. If the computer available for the modeling is less powerful than specified, a number of tricks that can be used:

- Reduce the number of dimensions.
- Use a steady-state model repeatedly for different time periods.
- Reduce the number of water quality parameters modeled.
- Reduce the number of calculation points or grid points (this can be done by nesting solutions, i.e., using a large, coarse grid first, then a smaller grid in the area of interest which uses the output of the coarse grid [see Hongzhou Bay and Chongqing Projects]). Use a less complex level in the model; many models have different levels of complexity and allow the user to select the level.
- Reduce the number of independent variables by removing the variable from the model or manipulating the coefficient associated with the variable (not possible in all models).

Many models are available at no cost. It is important that a good user's manual be available. User's manuals typically cost US$100.00. The model user should be competent in operating numerical models, understand receiving water processes, and have a university science degree or equivalent experience. The hydrodynamic and hydrological models are characterized by large time series input data sets. The user will need experience with data management, because calibrating the hydrodynamic model is normally the biggest and most complicated part of the model application. Good water quality predictions are based on good hydrodynamic predictions.

Many of the models are friendly and can be used by personnel with little experience in modeling. The user can learn by using the model in a progressive manner, from the simpler to more complex forms of the model, provided that the user understands the model equations. Model calibration is a trial-and-error process that can be simplified with an understanding of the model equations and some sensitivity analysis. Once calibrated, the model can be applied and, providing that the user quantifies the precision and accuracy of the model prediction, the predictions should be useful. A useful order check on the model predictions is to carry out simple mass balances of some of the substances on a spreadsheet.

The costs of water quality modeling are generally small, seldom more than a few hundred thousand dollars, including the collection of site-specific data. In fact, the modeling results may influence the design of the capital works. **In nearly all cases where water quality models are used, the costs of the modeling process are a small fraction of the savings on the capital works.** The largest variable in the modeling costs is normally the data collection cost. The costs for large dam projects where the modeling must define the operating characteristics for the dam as well as manage water quality in the reservoir tend to be the largest. Once the model has been developed, calibrated, and verified, it can be used for very little cost as a water quality management instrument. The model can be used for future project assessments as required, for example, to assess the merits of improved treatment for some of the discharges, or to quantify the impact of changing receiving water flows or background concentrations.

WATER QUALITY PARAMETERS

The most commonly predicted water quality parameters are discussed in this section of the guide, in particular why these parameters are modeled, what is modeled, and the reliability of the modeling process. Typical receiving water objectives are presented for each parameter, primarily to provide general information on the typical numbers that are in use; in any given project, the relevant receiving water quality objectives of the local regulating agencies must be used. A summary of the parameters, and methods used to measure these parameters, is presented in Table 2.1.

Dissolved oxygen. DO is required for most aquatic life and is one of the most important receiving quality parameters. Typically, fish like DO concentrations of between 5 and 8 mg/L. The generally accepted objectives for DO concentrations are instantaneous and/or seasonal averaged concentrations for rivers, lakes, and marine environments. In most instances, a minimum concentration (normally 3 to 4 mg/L) and a desired concentration (5 to 7 mg/L) are specified. **DO can be measured with a precision and accuracy of less than 5 percent.**

The DO kinetics in a natural water body is complex. Figure 2.1 shows the dissolved oxygen sources (external supply, photosynthesis, surface re-aeration, denitrification) and sinks (BOD), water column and sediment oxygen demand (SOD),respiration, and nitrification). Most of these processes are biological and occur over

Table 2.1 Water Quality Parameters Discussed in This Manual

Parameter[1]	*Environmental Properties*	*Modeling Considerations*
Dissolved oxygen	Required for aquatic life	BOD, SOD, nitrogens, photosynthesis & respiration, temperature, salinity, suspended solids
Nutrients - nitrogen	Required for aquatic plants; ammonia toxic to fish but preferred nutrient for most bacteria; dissolved oxygen sink and source	DO processes, aquatic plant demand, temperature, bacterial biomass
Nutrients - phosphorus	Required for aquatic plants	Aquatic plant demand
Indicator bacteria	Human intestinal bacteria	Salinity, temperature, suspended solids, sunlight
Suspended solids	Aquatic plants, media for bacteria & metals, asthetics, dissolved oxygen sink	Currents & bottom shears, partitioning coefficents
Heavy metals	Toxic to aquatic biota	Suspended solids transport, partitioning coefficients, pH, currents
Dissolved substances	Density, irrigation, aquatic biota	Currents, dispersion
Acidity	Toxicity, solubility of metals	Ionic balances
Temperature	Bacteria activity, fish habitat, solubility of oxygen, density	Heat flux balances
Oils, grease, PAHs	Toxic to aquatic biota, dissolved oxygen sink	Water surface phenomena, currents, winds

[1]There are numerous standard methods references like Standard Methods for the Examination of Water & Wastewater, American Public Health Administration; British Standards, BS 6068, Water Quality Part 2; German Industrial Standard Methods for the Examination of Water, Wastewater and Sludge, DIN 38404 to 38409; International Standards Organization Water Quality Standard Method; etc.

a period of time; the biological process rate is very sensitive to temperature.

DO is modeled as oxygen deficit, or the difference between the DO concentration and the concentration of the DO when the water is saturated with DO. The saturation concentration is a function of both the temperature and the salinity. Saturation concentrations are normally available in the model manual or other standard references (APHA, 1998).

The DO concentration is a function of numerous physical and biochemical processes. DO processes modeled are those shown in Figure 2.1. The model user has the option of omitting processes that may not be important in a particular application. For example, SOD is not important for rocky substrates, and photosynthesis and respiration are normally not factors in fast-moving rivers or receiving waters with high turbidities. **Many of the processes in the DO equation are difficult to measure; consequently, these processes are imprecise**. For example, the precision of the biochemical oxygen demand is 10-20 percent; re-aeration, 20-30 percent; photosynthesis and respiration, 10-20 percent; and sediment oxygen demand, 10-20 percent. **Simplifying the model by omitting processes that are known to be very imprecise can improve the precision of the model predictions.**

Nutrients. In water quality studies, only the vegetation nutrients nitrogen and phosphorus are considered because domestic wastewaters have high concentrations of nitrogen and can have high concentrations of phosphorus. Aquatic plants are part of the DO processes shown in Figure 2.1. Phytoplankton (microscopic free-floating plants) are the foundation of the aquatic biota in the receiving water as the food supply for zooplankton. **Without nutrients, aquatic organisms cannot exist; however, an excess of phytoplankton biomass can cause receiving water quality to degrade, primarily in the oxygen demands for the decay of expired phytoplankton bio-mass.** The ideal molecular ratio for phosphorus:nitrogen:carbon for plant growth is about 1:10:40. Phosphorus, nitrogen, and carbon can be measured with a precision of <10 percent. In general, aquatic plants in estuaries tend to be nitrogen-limited. High concentrations of nutrients, particularly nitrogen, can result in nuisance growths of aquatic plants and species. The nitrogen processes in the receiving water are shown in Figure 2.2, and the phosphorus processes in Figure 2.3.

Figure 2.1 Dissolved Oxygen Process

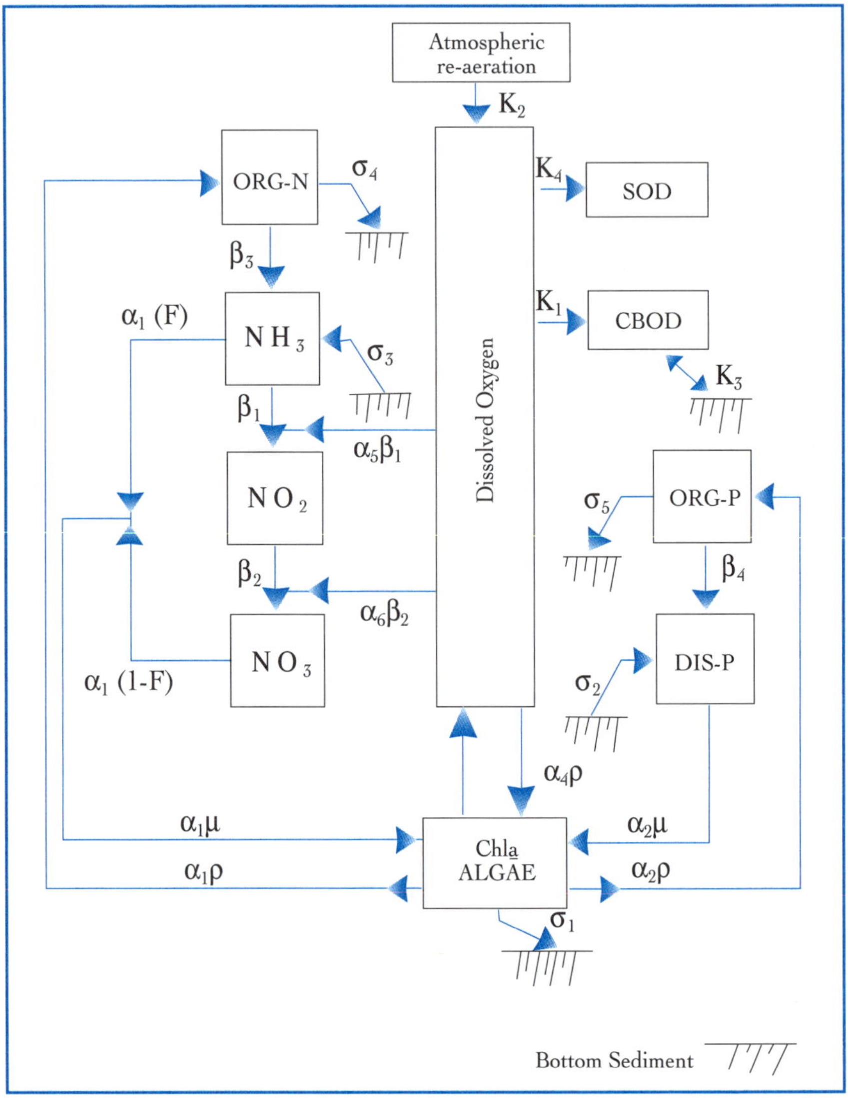

Nitrogen and phosphorus can be in both dissolved and solid forms in the receiving water and in both organic and inorganic forms. Complex models predict the concentrations in the various forms using partitioning coefficients, but only the dissolved forms

Figure 2.2 Nitrogen Processes

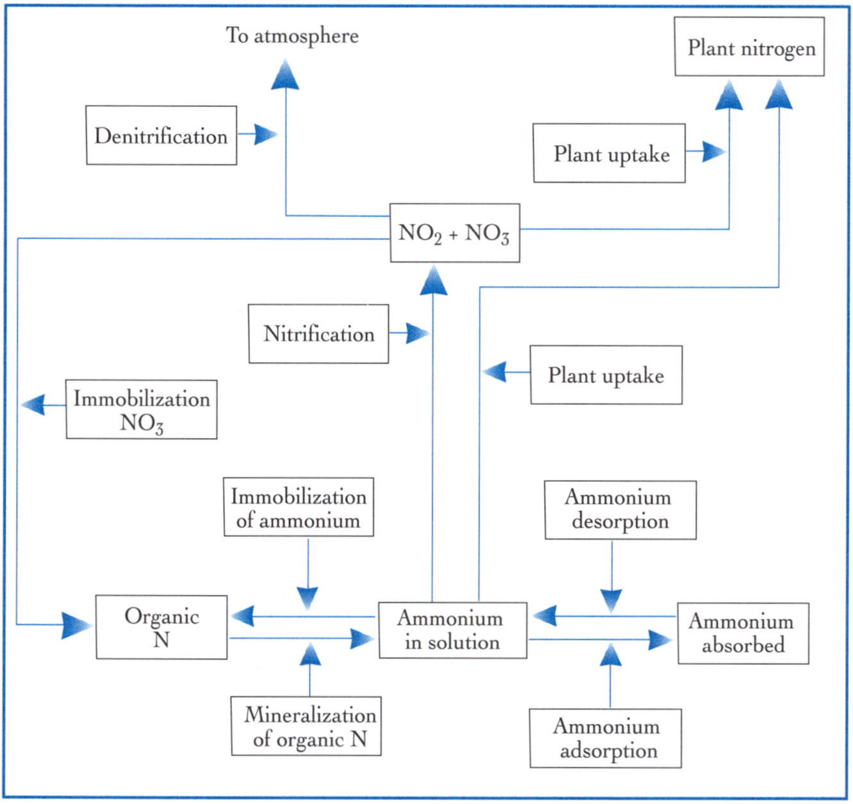

are readily available to aquatic plants in the short term. Many dynamic prediction models consider only the dissolved forms.

Nitrogen, in addition to being a nutrient of concern for aquatic plants, is toxic to fish in its ammonia form, and ammonia is the preferred nutrient for micro-organisms. Most regulating agencies use a receiving water objective of 0.02 mg/L of un-ionized ammonia, although some use the rainbow trout LC_{50} toxicity concentration of 0.08 mg/L as the maximum and 10 percent of the toxicity concentration for the long-term average. The portion of the total ammonia that is un-ionized is a function of the receiving water pH, water temperature, and salinity (APHA, 1996). The un-ionized portion is very high in basic waters (high pH) at high temperatures and vice-versa. The water quality model implications are that if ammonia toxicity is

Figure 2.3 Phosphorus Processes

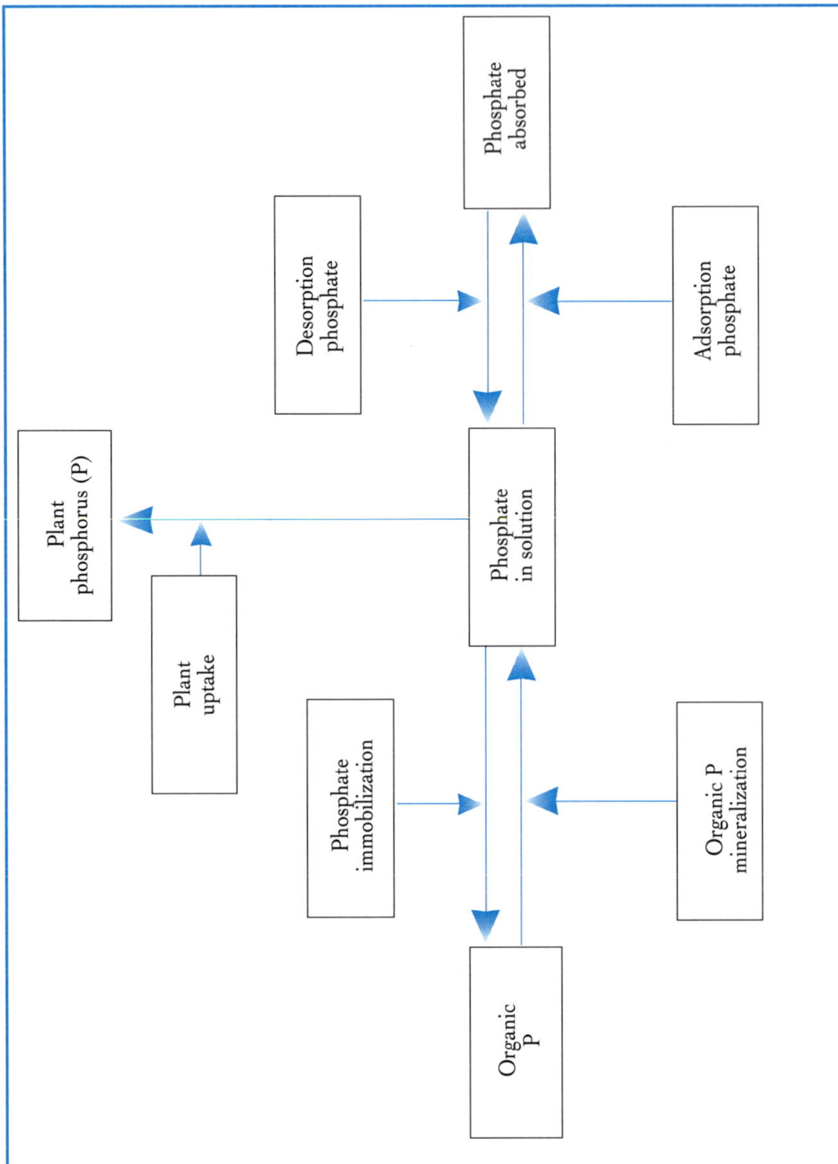

to be predicted, it is necessary to predict the pH and temperature and salinity in marine environments as well as total ammonia concentrations.

pH is difficult to model and predict. In most instances, pH must be measured in the receiving water, and statistical values for the pH are used to determine the un-ionized portion of the ammonia. The period that most affects the ammonia prediction is when the receiving water temperatures and pH are the highest and when the salinity is the lowest. Furthermore, because many agencies specify different concentrations for instantaneous prediction and averages over a period of days, appropriate time-averaged concentrations must be predicted.

Phosphorus is an aquatic plant nutrient. In natural freshwater receiving waters, phosphorus is frequently the nutrient that limits excessive aquatic plant growths. Domestic wastewaters are a source of phosphorus for the receiving water and can cause excessive aquatic plant growths, which will result in a degraded water quality.

Predicting the nutrient concentrations in receiving water using water quality models requires that the phytoplankton bio-mass and species as well as the macrophyte bio-mass be known. Because of the patchiness of phytoplankton bio-mass and the kinetics of the biomass (Palmer, 1981: Harris, 1980: Denman et al., 1977), quantifying it requires an extensive field measurement program. Furthermore, using bio-mass measurements in the water quality prediction models introduces more imprecision.

Indicator bacteria. The density of indicator bacteria is used as a measure of the presence of domestic wastewater in the receiving water and, consequently, the public health risk associated with human contact with these waters. The presence of domestic wastewaters has historically been determined by measuring a group of intestinal bacteria. Originally, a total coliform determination was developed. This test was subsequently replaced with a fecal coliform and *E. coli* test that was found to be more specific for the presence of intestinal bacteria. "Although *E. coli* is undisputed as the fecal indicator of choice, some of the fecal coliform tests used will also enumerate *Klebsiella* spp fecal sources" (NHW, 1992). Numerous attempts to define the bacterial density objective using epidemiological studies have met with with mixed results (see NHW, 1992 for a good discussion of bacteria indicators). Most regulating agencies use geometric mean density objectives of between 100 cfu/100mL (cfu = colony forming units) to 500 cfu/100mL (about 1 part sewage: 100,000 parts water) for recreational waters and for drinking water 0 cfu/100mL fecal coliforms and < 10 cfu/100mL total coliforms.

For recreational waters, model predictions must be suitable for the applicable receiving water quality objective, typically a geometric mean of 5 to 10 samples. The precision and accuracy of sampling, transporting, and measuring bacterial densities are about $\log_{10}$ 0.2 to $\log_{10}$ 0.35

In receiving waters, the bacterial cells are normally attached to organic flocs or suspended solids (see, for example, Palmer & Burrelle, 1996); consequently, the densities can vary between samples. Furthermore, these intestinal bacteria cannot survive for very long in natural water bodies, particularly if the salinities are high. The mortality rate of the bacterial cells is a function of many factors and generally not universal. The mortality rate is site-specific, and typically a 90 percent mortality occurs in 3 to 6 hours in marine waters (Palmer & Dewey, 1984) and in 12 to 30 hours in fresh waters. Mortality rates are normally measured by releasing and tracking a bacterial seed, frequently mixed with a visible dye tracer, either directly in the receiving water or in containment bags in the receiving water.

Most recreational indicator bacteria objectives are defined as the geometric mean of 5 to 10 samples over a period of time. Regulating agencies also frequently specify the field sampling methods to reduce variability in the samples. The model predictions should be geometric mean densities, and models that predict organic floc and suspended solids transport are better for indicator bacteria. The prediction models for indicator bacteria normally consist of transport, dispersion, and mortality. Transport and dispersion can be predicted with precision on the order of 50 percent, provided that the receiving water is not very dynamic. The mortality rate can cause the predictions to be imprecise, particularly in salt water receiving waters. In situ measurements of mortality are difficult unless a large bacterial source exists (Gannon and Busse, 1989; Palmer, 1984: and Springthorpe et al., 1993). The method for determining the densities of the indicator bacteria by culturing also introduces imprecision into the measurements, which affects the prediction model calibration and verification. Typically, the log standard deviation for the densities of indicator bacteria in the laboratory for split (three) samples is between 0.15 and 0.25 (Palmer, 2000). **An extensive quality assurance and control program is necessary for indicator bacteria model predictions.**

Suspended solids. Suspended solids can be inorganic or organic. Both are important in determining whether aquatic plant productivity and/or aesthetics will be affected. Most of the heavy metals like copper, zinc, lead, cadmium, and iron are associated with the very fine (cohesive) sediments. Indicator bacteria attach themselves to suspended solids, particularly the organic flocs. Similarly, organic compounds like pesticides and herbicides tend to combine with organic flocs. **The transport and fate of suspended solids are important in determining the concentrations of metals, bacteria, and organic substances in the receiving water.**

Suspended solids measurement is a mass measurement and as such generally underestimates the importance of the organic flocs and plankton bio-masses. The turbidity measurement, which is a light transmittance measurement, is frequently used in receiving waters. The interpretation of Secchi disc data is extremely difficult (Effler, 1985) and probably should be avoided in quantitative interpretations. There is no rigorous relationship between suspended solids and turbidity, although there is a general correspondence unless the plankton or organic floc bio-masses are large.

Domestic wastewaters have high concentrations of suspended solids. Typically, raw wastewaters have suspended solids concentrations of 100 to 200 mg/L (Metcalf & Eddy, 1992), which treatment systems can reduce to 10 to 50 mg/L. Receiving water objectives are typically expressed as a percentage increase (5 to 10 percent) over naturally existing conditions. For instance, the characteristics of the organic flocs are functions of the floc size, which is determined by the physical shear processes in the receiving water. Physical shear is difficult to predict in receiving waters. In general, the better sediment dynamics models use sedimentation velocity, resuspension velocity, and transport velocities, and **it is extremely important that site-specific field data be used in the predictions to improve their precision. Predicting the receiving water suspended solids concentrations and the suspended solids sedimentation is important**

in a model; however, predicting these processes for the very small cohesive particles and the organic flocs is difficult and requires complex models.

Metals. Metals are associated primarily with the suspended solids. **High concentrations of metals are toxic to aquatic life, with the dissolved form being the most toxic.** The metals of greatest concern are copper, cadmium, zinc, lead, mercury, cyanide, chromium, nickel, arsenic, and iron, with the lowest concentration objectives for mercury at 0.0002 mg/L, for cadmium at 0.004 mg/L, for copper at 0.015 mg/L, and for cyanides at 0.005 mg/L. Normally, the concentrations in domestic sewage are small and of little concern; however, if small industries such as metal shops, plating shops, and paint shops are connected to the domestic sewage system, metals can be a concern in domestic sewage. **Because the metals are associated with suspended solids, these metals can accumulate in the receiving water sediments and have a negative long-term effect on the receiving water and should be predicted in water quality models. Water quality prediction models must effectively predict the suspended solids transport and fate processes if metal concentrations are to be predicted in the receiving water.**

Dissolved substances. Dissolved nutrients and metals have been discussed in the preceding text. Other dissolved substances of interest in the receiving water are salinity, hardness, and salts. The total dissolved solids concentration is measured as conductivity with an accuracy and precision of 1 to 2 percent, and because temperature affects the conductivity measurement, the conductivity is given for a standard temperature. The mass of the various ions and cations can also be measured with an accuracy of 4 to 5 percent.

Large changes in the natural salinity of the receiving water cannot be tolerated by the existing biota. Consequently, regulating agencies generally restrict changes in salinity to less than 10 percent of existing levels. In marine environments, natural salinity is seasonal; therefore, model predictions should be for the season of low salinity. Salinity of more than 125 mg/L is also a concern in freshwater receiving waters if the water is to be used in irrigation. Sodium of more than 20 mg/L is also a concern for irrigation and drinking water. Hardness is a concern because it affects the toxicity levels for fish if metals and unionized ammonia are elevated in the receiving water.

Dissolved substances in the receiving water are diluted by bulk mixing and dispersion; consequently, these concentrations can be predicted well with most water quality models. The precision and accuracy for the total dissolved solids concentration predictions are similar to the precision and accuracy for the transport and dilution models. Conductivity data are useful in estuaries where the conductivity is on the order of 32,000 umhos/cm for salt water, municipal effluent is at 200 to 1000 umhos/cm, and the fresh water has a nearly zero conductivity level. Conductivity is used extensively in estuaries to quantify the fresh water fraction at any location in the estuary.

Total dissolved solids are measured as conductivity with an accuracy and precision of about 1 to 2 percent; however, conductivity is a function of temperature and must be given for standard temperatures. **In modeling, conductivity data are useful for mass balances in estuaries and in the vicinity of wastewater discharges.**

Acidity. **The pH in the receiving water is an important parameter for chemical reactions and toxicity. Acidic pHs permit metals to go into solution.** The pH is also important in the bicarbonate equilibrium in fresh waters, and it affects the levels of most toxicants for fish. Generally, the higher the pH, the less toxic toxicants are to fish. **pH is not predicted by most water quality models and must be measured in the field.** This can be done with a precision of 0.1 using standard pH meters.

Temperature. **Temperature is an important factor in the chemical reactions and biological activity in the receiving water**. Temperature is also an important parameter for fish habitat where the water temperature maximums and temperature ranges must be within specified limits for specified seasons. **The seasonally acceptable range and limits are defined by the by life cycle of the different fish species. Water temperatures are predicted well by many water quality models and can be easily measured with a precision of about 0.1^{0}C.** In estuaries, it may be necessary to measure the temperature with greater precision using more sophisticated equipment. **Existing water quality models are capable of predicting water temperature with an accuracy and precision of about 3 to 5 percent.**

Oils and grease. Municipal wastewaters are a source of oils and grease. Most regulating agencies specify that surface grease and oils be undetectable by sight or smell (<0.1 mg/L). Both are present in most domestic sewage and must be removed by treatment. Oils and grease are not predicted in most water quality models but can be predicted using the oil spill prediction models. **Measurements of oil and grease concentrations have a precision on the order of 15 to 20 percent.** For the oil spill models to be precise, extensive site-specific data are required. Seldom are these data available; consequently, **oil spill models typically have a precision of 25 to 50 percent.**

Polycyclic aromatic hydrocarbons (PAHs). PAHs are toxic to aquatic life at low concentrations (0.1 to 10 ug/L). **In most instances, the hydrocarbons are buoyant in the receiving water and require special oil spill or surface film water quality models that have a precision rate of 25 to 50 percent (unless extensive site-specific data are available).** Hydrocarbons that are associated with suspended solids can be predicted using water quality models that predict the suspended sediment transport and fate. Any model will have difficulty predicting PAH concentrations at the low objective levels; consequently, PAHs can only be predicted qualitatively.

RECEIVING WATER PROCESSES

When a substance is discharged to a receiving water, many different receiving water processes move the substance, dilute it, and change it. These processes are the basis of all water quality prediction models. Some processes are easier to model than others, and the processes are not independent of each other. One of the major strengths of water quality models is that the model integrates the interaction of these processes.

Physical processes. Dilution is simply the mixing of a volume at one concentration with another volume at a lower concentration. Simple batch computation produces the resulting concentration. For continuous flows, batch computation can be done over a period of time. This computation is frequently carried out as a first-order approximation or as a means of confirming more complex model predictions. It is sometimes called a "reality check."

Dispersion refers to the mixing processes that occur as a result of eddies within the fluid, differences in velocity in different areas in the fluid (sometimes called the shear), and gradients of concentration. Dispersion is normally expressed as a temporal constant (and frequently as a spatial constant) associated with the concentration gradient terms in the mass balance equation. In fact, the dispersion coefficient varies with time (Alsaffar, 1966; Okubo, 1971), so the most appropriate dispersion coefficient is selected for the model. Dispersion coefficients in a water quality model are frequently adjusted in the calibration of the model. For some unexplained reason, the model calibration dispersion coefficient may be very different from site-specific field measurements. In these situations, the model calibrated dispersion coefficient should be used in the prediction process.

Advection is the (velocity multiplied by concentration) term in the mass balance equation. It is generally referred to as the mass transport in the direction of the velocity. In models, advection is a complex process because both the velocity and concentration are time and spatial variables.

Tides affect water levels, which changes the volume of water; therefore, there is more dilution water during high tides and less during low tides. But there are other considerations for a freshwater waste discharge to marine water. The difference in densities between the freshwater effluent and receiving water reduces the rate of mixing of the two waters. If freshwater sources exist in the coastal region, salinity gradients can exist; when that is the case, the differences in water density can generate density-difference-driven circulation in the coastal waters. In long estuaries, the tidal effects on water quality can be very complex because of the temporal, spatial, and seasonal variability of the tidal processes combined with the variability of discharges.

Most **river flows** vary seasonally and can have shorter-period variations associated with rainfall. River flow is an important water quality parameter that most water quality models treat as a spatial and temporal variable.

Currents in the receiving water can be generated by winds, tides, density, thermal differences, or river flows. These currents and their spatial and temporal variations produce dilution and dispersion in the receiving water. In a narrow river, the currents are directly related to the river flow, but in wide rivers, lakes, estuaries, and marine coastal regions, the currents are much more complex. Currents are vectors (speed and direction) which normally vary with time, location, and season.

Buoyancy, both positive and negative, is the difference in density between the effluent and the receiving water. The largest differences in density occur for freshwater discharges into a salt water receiving water; however, the density differences between effluents at temperatures different from the freshwater receiving water are also a concern. The density differences reduce the rate of the mixing of the effluent and the receiving water, resulting in higher concentrations in the receiving waters.

Density stratification in receiving waters is a seasonal phenomenon in lakes, reservoirs, and marine coastal regions and is a permanent feature in estuaries. The density depth stratification reduces the rate of mixing with depth and changes the physical transport processes in the various depth layers of the receiving water. Receiving water depth stratification is an important factor in determining water quality in the receiving water and, if it exists, must be addressed in the water quality models. In general, three-dimensional models are avoided because of the difficulty in calibrating these models. Specialized models like the oil spill models have been developed for buoyant contaminants in the receiving water, and some two-dimensional models can be used in the depth for density stratification and in the downstream dimensions.

Suspended sediment kinetics in the receiving water is the main mechanism for the transport of bacteria, heavy metals, and organic contaminants. Both the organic flocs and inorganic suspended cohesive solids are important. A water quality model for sediment kinetics should include settling, transport, resuspension, and partitioning. Sedimentation velocity is a function of the particle size and density difference between the particle and the receiving water. Resuspension is a function of the physical shear at the bottom, type of bottom, and particle size defined by mobility and transport parameters (Ackers & White, 1973; Yalin, 1977; Van Der Kooij et al., 1991). Partitioning coefficients that identify the portion of the metal in both the solid and dissolved states are different for the different metals, concentrations of suspended solids, and their porosity. Sediment kinetics is so complex that few water quality models address them in a comprehensive manner. Most models address only the sedimentation and partitioning processes.

Surface processes. **The most important surface process for general water quality modeling is re-aeration** because the two main mechanisms for supplying oxygen to receiving water are surface re-aeration and photosynthesis. Surface re-aeration in receiving waters is a complex process that is a function of the surface area, turbulence, wind climates, dissolved oxygen concentrations near the water surface, and temperatures of the air and fluid. Surface re-aeration can be measured by injecting a gas like propane into the receiving water and measuring the escape of this gas. However, because of the cost of these measurements and the number of measurements required for all the different conditions, re-aeration is normally determined by using empirical relationships based primarily on fluid velocity and depth of the receiving water. Most models allow the user to select the relationship from a range of options. Frequently, re-aeration is used as a calibration variable.

Volatilization refers to the changing of a substance from a liquid to gaseous state at the liquid interface. The process is a function of Henry's Law, transfer coefficients and concentrations (Lewis & Whitman, 1924; Gowda & Lock, 1985). **Volatilization is an important process for organic contaminants of concern.** For general water quality modeling, the main concerns are PAHs, and their fate in the receiving water is approximated by a single factor in the oil spill models.

Atmospheric deposition is an important process for nutrients, particularly for phosphorus in freshwater lakes. Most water quality models do not consider atmospheric deposition as a process in the model, though it may be included in models as a non-point source using locally measured deposition rates per area or literature deposition rates.

Biochemical processes. Chemical processes occur during a short period of time, whereas biological processes occur over a much longer period.

Dissolved oxygen is essential for most aquatic biological life. DO processes are designated as either sources or sinks which when summed result in the DO concentrations in the receiving water. These processes are shown in Figure 2.1. **Because all of the DO processes are biological except for surface re-aeration, time is required for them to take place.** The rates of these processes are also functions of temperature. In general, biological activity increases with temperature. One of the DO sinks is the biochemical oxygen demand, which is measured over a five-day period according to methods in various publications (APHA, 1996; ASTM *Annual vols. 11.01 & 11.02*; British Standards, British Standards Institute, *Water Quality BS-6068*). When large amounts of BOD are discharged into a receiving water with a long residence time, it is often necessary to measure BOD for periods longer than five days, referred to as "the ultimate BOD" (National Council, 1982, 1986). Atmospheric re-aeration and sediment oxygen demand (Figure 2.1) are difficult to measure in the field, and empirical methods are frequently used in models for these processes. The empirical methods for re-aeration are based on flow characteristics and depth, and for sediment oxygen demand are based on the organic carbon concentrations in the sediments. Using these empirical methods in water quality models affects the precision and accuracy of the model predictions.

Phosphorus is a necessary nutrient for the growth of aquatic flora (Figures 2.1 and 2.2). The natural sources of phosphorus are generally limited; consequently, natural freshwater systems often have little phosphorus. Domestic sewage, on the other hand, often contains large amounts of phosphorus, and discharges of untreated domestic sewage can change the nutrient balance in the receiving water. As aquatic flora require only small amounts of phosphorus, **treatment systems must remove the phosphorus to concentrations to about one mg/L to control the eutrophication in the receiving water.**

Nitrogen is a necessary nutrient for the growth of aquatic plants (Figures 2.1 and 2.3). Domestic sewage (typically 25 to 30 mg/L of total organic nitrogen) and animal wastes contain high concentrations of organic nitrogen, which can cause eutrophication in the receiving water. (The ideal molecular weight ratio of nitrogen to phosphorus is about 10:1; therefore, the receiving water is more sensitive to phosphorus discharges than to nitrogen discharges.) However, because much of the nitrogen is in an ammonia form, it exerts a high oxygen demand on the receiving water. Furthermore, high concentrations of ammonia are toxic to fish.

Photosynthesis and respiration of aquatic flora are important in DO balance in the receiving water. The photosynthesis and respiration processes are shown in Figure 2.1. In the water quality models, the phytoplankton growth rates are expressed in equations that have terms for light, extinction coefficients, depth, temperature, and carbon:chlorophyll a ratios. Another approach to quantifying photosynthesis and respiration is to measure DO concentrations and temperature at a fixed point for a period of 30 hours. Data collected from these measurements can be used to quantify photosynthesis and respiration.

Bio-mass transfers in the aquatic food web system, including detritus. Most natural receiving waters contain suspended solids, which consist of inorganic particles—principally clay particles, phytoplankton and zooplankton, and organic flocs—referred to as detritus. The detritus is consumed by zooplankton and fish and ultimately returned to the receiving water as a waste product. **The detritus may have high bacterial densities, heavy metal, and other organic compound concentrations** which may be returned to the receiving water or remain in the zooplankton and fish. Bio-mass transfers in this manner can be important when predicting the concentrations of water quality parameters in the receiving water.

Indicator bacteria are intestinal bacteria from warm-blooded animals; they cannot normally survive in natural receiving waters. Their mortality rate in the receiving water is referred to as T90, which is the time required for an order of magnitude reduction in the density of the bacteria. In a marine receiving water, the T90 is typically 3 to 6 hours; in freshwaters it is 12 to 30 hours. The T90 is function of temperature, particle association, sunlight, and other factors. T90 can be measured in the field using various methods, but in most instances, literature T90s or a range of T90s are used in water quality models with some site verification data. Sometimes it is best to select a T90 value that is too large so that the predictions will be conservative.

Sediment water interactions are important for the water quality parameters that are associated with suspended solids like indicator bacteria, heavy metals, and organic substances of concern. The portion of the water quality parameter in a dissolved form and associated with suspended matter is generally defined by the partition coefficient. This coefficient is a characteristic of the substance being modeled. Most water quality models predict total concentrations, which consist of the dissolved and suspended solids components, with the portion in each state determined from the partition coefficient.

General Prediction Confidence

The physical and biochemical processes discussed in the preceding sections can be measured and modeled with varying degrees of confidence. It is helpful to list what can be measured in the field and what is difficult to measure. Water quality parameters that can be easily measured in the field with some precision are:

- physical parameters like position, depth, temperature, salinity, turbidity, particle size, and sediment-settling velocities;
- chemical parameters like concentrations of most substances, including dissolved oxygen and suspended solids, and pH; and
- biological parameters like bacterial densities, phytoplankton and zooplankton bio-masses, and fish species and populations.

Water quality parameters that are difficult to measure in the field are:

- physical parameters like bottom roughness, circulation patterns, resuspension of sediments, surface re-aeration, partition coefficients, transport and fate of organic flocs and cohesive sediments, and atmospheric loadings;
- chemical parameters like sediment oxygen demand and volatilization; and
- biological processes like indicator bacteria mortality and detritus biomass recycling.

In most instances, processes that are difficult to measure are defined by empirical methods or through literature values for similar receiving water. If empirical methods or literature values are used, the model prediction precision is affected. The process of defining the precision and accuracy of models is based on the quality assurance and quality control in both the field data collection and analysis and modeling process.

CHAPTER 3

Some Commonly Used Models

An overview of water quality models was presented in Chapter 1. There is a general structure in the water quality models being used today. This structure is discussed in this chapter. Understanding this structure will assist a potential model user in evaluating the characteristics of any model. Most of the models have three parts, which are discussed in the following text. The model user in many cases can omit processes that may not be important in a particular application. These simplifications are discussed. Equations are presented to show the required user inputs to the model for the different processes in the receiving water. While it is true that every model has some unique characteristics, a general common structure exists in the models. This common structure consists of three parts: 1) the hydrodynamic/hydrological part, 2) the mass balance part, and 3) the receiving water process part. Much of the following discussion is based on the information contained in the various model manuals that are discussed in the Appendix.

HYDRODYNAMIC MODEL

The hydrodynamic characteristics, namely the spatially and temporally varying velocity vectors and water levels, can be determined by solving the following equations, shown below.

Receiving Water Equation of Motion

The equation represents the change of local inertia and rate of momentum change.

$$\frac{\partial U_i}{\partial t} = -U_i \frac{\partial U_i}{\partial x_i} + \sum a_i \qquad (1)$$

where U_i = velocity in the i direction
t = time
x_i = distance in the i direction
a = gravity, friction, and wind acceleration

$$Gravity = -g \frac{\partial H}{\partial x_i} \qquad (2)$$

where H = depth
g = acceleration resulting from gravity

$$Friction = -g \frac{n^2}{R^{1.33}} \; (U_i | U_i |) \qquad (3)$$

where n = bottom friction
R = hydraulic radius = wetted area/perimeter

$$Wind = \frac{C_\partial}{R} \frac{\rho_a}{\rho_q} W^2 \cos \Phi \qquad (4)$$

where C_∂ = surface drag coefficient
ρ_a, ρ_w = density of air and water
W = wind velocity at 10 m
Φ = wind angle

Receiving Water Equation of Continuity

The continuity equation is the time-varying water mass balance relationship, including water depth.

$$\frac{\partial H}{\partial t} = -\frac{1}{B}\frac{\partial Q}{\partial x_i} \tag{5}$$

where B = width

Q = flow

The unknowns are the velocities and depths at various locations and times. To solve the equations for these values, it is necessary to use numerical methods on a spatial grid or elements. The user is required to define the spatial grid, the time step for the numerical solution, the upstream and downstream boundary conditions as functions of time, the initial conditions, element cross-sectional information, and values for n and C_∂. The values for n and C_∂ are estimated; then the model is used. The predicted depths and velocities are compared to the values in the calibration data set. If the H values are too high, n is reduced and the procedure is repeated until the H simulated values match the calibration data set H values. Next, the velocities are adjusted to match measured values by adjusting C_∂. The calibration is a trial-and-error process that can be tedious, particularly when verification data sets are also used, requiring further adjustments to the model.

This process is simplest in one dimension, becoming progressively more difficult in two and three dimensions. Primarily, n is adjusted in the calibration process, and sometimes depth is adjusted to ensure that water is not accumulating or running out of the segment for the modeling period. The adjustments to C_∂ are normally minor. Theoretically, both n and C_∂ are probably different for each element in the model; however, to do this in the calibration process would be very time-consuming. In a typical model, n would have 5 to 10 values over the modeling grid.

There is some numerical dispersion (E_{num}) precision introduced by the numerical solution (backward or central differencing or other schemes) which is a function of the time step (Δt), spatial grid size (L), and velocity (U) ($E_{num} = (U/2)(L - U\Delta t)$). Many manuals provide methods for determining the numerical dispersion for the model numerical solution used, as well as methods for applying a factor to the advection terms which will reduce the numerical dispersion on the predictions. And because the model predictions are for grid locations and "n" and "C_d" are assumed constants for areas of the model and time, the predictions can be expected only to match measured data sets approximately. The velocities and water depths predicted from these two equations are used as inputs to the next model part.

The only model simplification possible for the hydrodynamic part of a model is to assume steady-state conditions and reduce the dimensions to two or, if possible, one. There are a couple of tricks that can extend the capabilities of simplified models. Steady-state models can be run repeatedly for different conditions to simulate time-variable conditions, and in some instances the model dimensions can be reduced to one dimension by using streamlines as an axis.

MASS BALANCE

Discharged Substance Mass Balance Equation

A general mass balance equation is the time-varying conservation of the mass of a substance dissolved or suspended in the water.

$$\frac{\partial C}{\partial t} = -\frac{\partial (U_i C)}{\partial xi} + \frac{\partial}{\partial xi}\left(E_i \frac{\partial C}{\partial xi}\right) + \sum S_j \tag{6}$$

where
C = concentration
U_i = velocity in direction i
∂x_i = distance in direction i
E_i = diffusion coefficient direction i
S_j = sources point and non-point, boundary loading rate, atmospheric, kinetic transforms

In the general mass balance equation above, the first term on the right-hand side of the equation is referred to as the advection or transport component, the second term is the dispersion component, and the last term is the sources and sinks.

The finite difference form of the mass balance equation for the numerical solution consists of the following.

Discharged Substance Transport Equations

Transport equations are used to represent the movement of a substance dissolved or suspended in the water.

$$\frac{\Delta V}{\Delta t} = \sum (flows + precipitation - evaporation) \quad (7)$$

where V = volume

$$\frac{\Delta(VC)}{\Delta_t} = \sum -QC + \sum -Q_p C f_D + \sum\sum -w_s A C f_s + \sum R\Delta C + \sum (R_p(f_D \Delta(C \div n))) + \sum W + \sum\sum V S_k \quad (8)$$

where
C = concentration
Q = flow
Q_p = pore water flow
f_D, f_S = dissolved and solids fractions
w_S = solids transport velocity
A = area
R = dispersive flow
R_p = dispersive pore water flow
W = sources and sinks – point, non-point boundary sources
S = kinetic transforms

Each parameter introduces another equation as shown below.

Pore Water Advection

$$\frac{\partial M_i}{\partial t} = Q_{pij} f_{Dj} C_j \div n_i \quad (9)$$

where M = mass of chemical

C = total chemical concentration
N = porosity
f = dissolved fraction of chemical
Q = pore water flow rate

Sediment Advection

$$H_i \frac{\partial Si}{\partial t} = w_D S_j - (w_R + w_S)\, S_i \tag{10}$$

where i = sediment
j = water
H = depth
s = sediment concentration
w_D = deposition velocity
w_R = scour velocity
w_S = sedimentation velocity in upper benthic layer

The user can select any or all of the advection relationships above. In all cases, the user must provide the segment interfacial areas, characteristic lengths, and segmentation. In addition, for the sediment advection, the sediment transport velocity and fraction absorbed to sediment must be provided.

Similar mathematical relationships can be developed for the dispersion terms. In these relationships, the user must provide the dispersion coefficient as a function of time and, for the pore water, the dissolved fractions in the water and sediment.

In the mass balance part of the model, the user can add or delete advection or dispersion terms to suit a particular application of the model. However, the addition of each term requires that the user define the appropriate coefficient for the model application. The next model part is the receiving water processes.

RECEIVING WATER PROCESSES

Dissolved Oxygen

The receiving water DO processes are shown in Figure 3.1. These processes can be expressed in an equation as follows:

$$\frac{\partial O}{\partial t} = K_2\,(O_{sat} - O) + (\alpha_3\mu - \alpha_4\rho)Gn - K_1L - K_4/H - \alpha_5\beta_1N_1 - \alpha_6\beta_2N_2 \quad (11)$$

where O, O_{sat} = DO and DO saturation concentration (mg/L)

a_3 = rate of oxygen production per unit of algal photosynthesis (mg$O$/mgGn)

a_4 = rate of oxygen uptake per unit of algal respired (mg$O$/mgGn)

a_5 = rate of oxygen uptake per unit of ammonia nitrogen (mg$O$/mg$N$)

a_6 = rate of oxygen uptake per unit of nitrite nitrogen (mg$O$/mg$N$)

m = algal growth rate (temperature dependent) (1/day)

r = algal respiration rate (temperature dependent)(1/day)

Gn = algal bio-mass concentration (mg/L)

H = depth (m)

L = concentration of ultimate BOD (mg/L)

K_1 = BOD deoxygenation rate (temperature dependent) (1/day)

K_2 = re-aeration rate (temperature dependent) (1/day)

K_4 = SOD (g/m^{2}day)

β_1 = ammonia oxidation rate coefficient (temperature dependent) (1/day)

β_2 = nitrite oxidation rate coefficient (temperature dependent) (1/day)

N_1 = ammonia nitrogen (mg/L)

N_2 = nitrite nitrogen (mg/L)

Equation 11 states that the dissolved oxygen concentration is the sum of the sources (re-aeration and net algal production) and the sinks (BOD, SOD, and nitrogen oxidation). Most models include algal growth equation options based on the available light and photosynthetic rates, which the user can select. If algal production is not a factor in the oxygen balance (e.g., if receiving water turbidity is high or is fast-running water or is nutrient-depleted or chlorophyll *a* <10 ug/L), the algal oxygen production term can be omitted. Some dissolved oxygen measurements over a 30-hour period during the growth period for aquatic plants can be used to determine whether algal bio-mass is a factor in the dissolved oxygen balance. Similarly, other terms in the equation can be omitted if these are not considered a factor. The terms can also be extended if necessary. For example, macrophytes may be the largest source of oxygen production. In this case, an area measurement term would have to be added for the macrophytes, that is like the SOD term, not a volume measurement like the algal bio-mass term.

As discussed previously, the model prediction precision is generally improved if the model is simplified. Re-aeration and SOD are difficult to measure in the field. Re-aeration is normally computed from empirical relationships for the type of receiving water (lake, river, and ocean). These empirical relationships are available as options in many models. Some DO depth profile measurements near the bottom will clearly show whether SOD is a factor. If it is, the DO concentrations will be lower just above the bottom sediments. These profiles should be measured when the receiving water is at its highest temperature. In general, if the total organic carbon in the sediments measured by the loss on ignition is less than 3 percent, SOD is probably not significant. If SOD is a factor, some in situ measurements should be made. It is also possible to quantify the SOD by a method of difference. In other words, provide all the other sources and sink information to the DO balance, then assign the difference to SOD; however, because the re-aeration as quantified by empirical equations can be imprecise, SOD should be measured if it is a factor in the oxygen demand balance.

Some models allow the user to specify the level of complexity to be used in the model. In the case of the DO balance, these levels may be as follows:

1. BOD and SOD
2. BOD (carbonaceous + nitrogenous) and SOD
3. Full equation.

Using the model at a lower level of complexity is a useful approach when the amount of site-specific data is limited. It is normally possible to determine whether a more complex level of modeling is required for a particular application by testing the simplified model on separate verification data sets. If the predictions from the simplified model differ from the verification data sets, more advanced forms of the model should be tried. In this way, the appropriate level of the model will be identified.

Nutrients

The nutrient processes are presented in Figures 2.2 and 2.3. The nitrogen can be considered to exist in four components: phytoplankton nitrogen, organic nitrogen, ammonia, and nitrate. Although some models lump some of these components together, the four will be discussed separately here. Nitrogen processes in the receiving water are complex, and considering the four nitrogen components separately simplifies the modeling process.

Ammonia (C_1)

$$\frac{\partial C_1}{\partial t} = \text{(mineralization)} - \text{(growth)} - \text{(nitrification)} + \text{(death)} \qquad (12)$$

Mineralization = conversion of organic nitrogen to the inorganic form

Growth = take-up of nitrogen by the phytoplankton.

Nitrification = conversion to nitrate.

Death = recycling of organic nitrogen from phytoplankton mortality

Nitrate (C_2)

$$\frac{\partial C_2}{\partial t} = \text{(nitrification)} - \text{(growth)} - \text{(denitrification)} \tag{13}$$

Denitrification = nitrate to nitrogen

Phytoplankton Nitrogen (C_3)

$$\frac{\partial C_3}{\partial t} = \text{(growth)} - \text{(death)} - \text{(settling)} \tag{14}$$

Organic Nitrogen (C_4)

$$\frac{\partial C_4}{\partial t} = \text{(death)} - \text{(mineralization)} - \text{(settling)} \tag{15}$$

There are obviously many coefficients, rate terms, and fraction partitioning required for the components of the nitrogen process. Manuals do provide a range of values for the required inputs and the default options that the model will use if the user does not provide values to the model.

Similarly, phosphorus kinetics can be considered as three components: phytoplankton phosphorus, organic phosphorus, and inorganic phosphorus (orthophosphate). The kinetics of these components can be represented by the following equations:

Phytoplankton phosphorus (C_5)

$$\frac{\partial C_5(P/C)}{\partial t} = \text{(growth)} - \text{(death)} - \text{(settling)} \tag{16}$$

where P/C = phosphorus to carbon ratio in phytoplankton

Organic phosphorus (C_6)

$$\frac{\partial C_6}{\partial t} = \text{(death)} - \text{(mineralization)} - \text{(settling)} \tag{17}$$

Inorganic phosphorus (C_7)

$$\frac{\partial C_7}{\partial t} = \text{(mineralization)} - \text{(growth)} - \text{(settling)} \tag{18}$$

Like the nitrogen component equations, coefficients, rate parameters, and partitioning are required for the phosphorus processes. The range of values for these required inputs is provided in the manuals, as well as the default options.

Some models allow the user to select the level of complexity for the phytoplankton-nutrient kinetics similar to the DO balance. If the data available for the site are limited, simpler models once again are more appropriate, at least initially.

Heavy Metals

Heavy metal kinetics in a receiving water is complex because the metals can exist as soluble organic or inorganic complexes, sorbed onto organic or inorganic particles, and precipitate or dissolve. All the soluble components can be lumped into the dissolved term. WASP4 provides a modeling framework at four levels of complexity. Because the partitioning coefficients depend on the sorbent character of the suspended solids, there are no consistent partitioning coefficients. Site-specific measurements are required for heavy metal predictions. The transport kinetics of suspended solids is included in the mass balance part of the model (see equation 10); however, the partitioning coefficients in this equation are for the liquid or solid stage. The partitioning of a substance between dissolved and sorbed for equation 10 is predicted in this model component. If site-specific data are limited at the site, simpler model configurations should be used. For example, in the WASP4 model, the user can select from the following levels of complexity for the metal predictions:

1. Specify average concentration field by setting the initial conditions. The solids concentrations will then influence the chemical partitioning.
2. Specify average concentration field and settling, deposition, scour, and sedimentation velocities.
3. Simulate total solids by specifying loads, boundary concentrations, and initial conditions, settling, deposition, scour, and sedimentation velocities.
4. Simulate three sediment types as in Level 3.

Heavy metals are associated primarily with the cohesive sediments, or organic flocs. In general, cohesive sediments will not settle if the velocity is greater than about 12 cm/sec, and resuspension occurs when the velocity is greater than 20 cm/sec. Knowing the critical velocities and the velocities in the receiving water, it may be possible to simplify the sediment dynamics model.

Temperature

Many of the coefficients, rate parameters, DO saturation concentration, and unionized portion of ammonia are temperature-dependent; therefore, temperature must be predicted for the receiving water. The generalized form of a temperature equation is as follows:

$$\frac{\partial T}{\partial t} = \frac{\partial\left(A_x E \frac{\partial T}{\partial x}\right)}{A_x \partial x} - \frac{\partial (A_x U T)}{A_x \partial x} + \frac{Q H_N}{\rho c H} \tag{19}$$

where T = temperature

A_x = cross-sectional area

E = dispersion coefficient

U = mean velocity

ρ = density

c = heat capacity

H = depth

QH_N = rate of heat input

= net short wave + net long wave – outgoing long wave back radiation flux ± conductive flux – evaporation heat loss

This particular form of the temperature prediction may be simplified for a particular application. Statistical methods may determine some simple relationships between the air temperature and water temperature in a receiving water. Another approach to simplify the modeling process is to use the maximum and minimum recorded temperatures in the receiving water to determine the range of values for the various coefficients. However, the complete temperature prediction equations are required for reservoirs or large thermal discharges to the receiving water.

Oils, Grease, and PAHs

These substances are buoyant and do not mix well with the receiving water; consequently, they remain on or near the water surface, where they spread outward as a thin surface film. Special models have been developed to predict the behavior of these surface films, which are referred to as oil slick models. Oil slick models are Lagrangian models that follow the path of the oil slick dispersing and diluting the oil slick in the receiving water. Like other water quality models, oil slick models require a velocity vector field. The hydrodynamic equation (equation 1) includes wind-generated currents (equation 4) and can be used to determine the surface current vectors, although these currents are depth-averaged in the model formulation. If the hydrodynamic predictions are not available, the surface current vectors can be estimated from wind data (3 ± 2 percent wind speed at 7 ± 6 degree deflection) (Huang and Monastero, 1982; Venkatesh, 1990). In the receiving water, the processes operating on the oil parcels are as follows:

- surface tension spreading normally early in the oil parcel release;

- dispersion – turbulence and physical spreading; and
- weathering – includes evaporation, depth dispersion, emulsification, dissolution, and biodegradation.

For periods of a few days, the slicks can be predicted well using only the time and spatial variable velocity field and dispersion data. The models consist of releasing individual parcels of oil and tracking the movement of the parcel of oil through the velocity field as it is moved by the currents and dispersion. The location of the parcel on the two-dimensional grid is determined at selected times after its release. Typically, 200 to 300 parcels of oil are released to obtain a representative statistical sample for the oil slick. The oil patch is then represented by a plot of the individual parcels. Statistical analysis of the parcels defines the mean concentration and variance at different times after release and for different distances from the start of the spill. The MIKE programs discussed in the appendix have an oil spill model.

Summary

Most mechanistic models consist of a hydrodynamic part, a mass balance part, and a receiving water process part. The hydrodynamic part predicts water levels and currents. The hydrodynamic equations must be solved numerically, which requires that the user provide boundary and initial conditions, bathymetry, time and/or spatial elements, wind data, bottom friction, and wind surface drag.

Hydrodynamic calibration is a trial-and-error procedure that may be tedious. The simplest form of the hydrodynamic model is the one-dimensional steady-state model (QUAL2). In some instances, this model can be used repeatedly to simulate different conditions at different times, and can be applied along streamlines in two- or three-dimensional flow fields.

The mass balance and process parts of the model use the outputs from the hydrodynamics part. The mass balance part transports and disperses substances and balances the discharges, input flows, and outflows. Besides providing the point and non-point discharges and other loadings as well as the initial conditions, the user must provide the dispersion coefficients and, for suspended solids, partitioning coefficients. The dispersion coefficients for the model are normally quantified in the calibration process. The receiving water process parts can be complex, requiring many different coefficients, rate parameters, and partitioning coefficients. Every effort should be made to simplify these processes for a particular model application.

Discharged substances that are both buoyant and that do not mix well with the receiving water (e.g., oils and PAHs) require a surface spill type of model.

SELECTED MODELS

In preceding text, the different processes were discussed to develop an understanding of the independent variables in the model prediction equations. It is not necessary that all these processes be included in a model for a particular application; nevertheless, the model user should know what processes have been omitted in the model and the rationale for not considering them. One of the reasons for omitting processes may be the lack of site-specific data and the reluctance to use literature or default values instead of the site-specific data. Or, the user may want to develop a better understanding of the receiving water responses by using a simplified version of the model to predict water quality, then compare the model predictions from different model formulations. For example, a user may use the same model to predict receiving water quality for two different loadings from an outfall or compare the receiving water quality predictions for an outfall at two different locations.

In many instances, it may be more efficient to use more than one model for a project or to combine parts of several models. If the receiving water processes and discharges are very complex, it is always easier for the user to understand the receiving water quality kinetics if the models are simplified. Some of the models can be used as a black box with little site-specific data inputs to the model (see Appendix). Because the processes discussed above have many site-specific user data input requirements, using the model as a black box should be avoided if possible. If it is necessary to use a black box model, it is important that the model user quantify the prediction precision if the model is used comparatively and both precision and accuracy if the model predictions are compared to receiving water quality objectives.

Some specific models are discussed in the Appendix. These models may not necessarily be the best models for any particular application. They do include some of the models that have been used for World Bank projects and some other well-used models that are readily available at no cost.

The World Bank Pollution Prevention and Abatement Handbook, 1998 discusses four representative models: QUAL2E, WASP, CE-QUAL-RIV1, and HEC-5Q. In this section, the models CE-QUAL-RIV1 and HEC-5Q are not discussed, although these models are similar to other models that are discussed. As discussed previously, there is a generic structure common to most of the models so that the individual models can be viewed as different ways of packaging the three parts of the model generally for a specific application. *The World Bank Pollution and Abatement Handbook*, 1998 uses a table to show the model characteristics, which includes the type of receiving water, time characteristics, and water quality parameters predicted. This model classification system is presented in Table 3.1 with some additions.

The models presented in the Appendix are either specialist models or general models. The general models can all be used as steady-state models or in one or two dimensions; therefore, these models can be used for river or lake or ocean receiving waters. The general models are also designed so that parts of the model can be used in other models; therefore, the user can create a hybrid model. Summaries of the important features and limitations of each model are presented here. As discussed previously, the water quality parameters of interest in this manual are temperature, turbidity, suspended solids, dissolved oxygen, nutrients (including ammonia), indicator bacteria, oils, grease, PAHs, and heavy metals. The water quality parameters that a model cannot predict are identified as a limitation in the model description.

Outfall Models – CORMIX (USEPA)

Most outfalls end in some kind of diffuser, which can be a single port or multiport diffuser.Diffusers increase the local dilution (commonly called the initial dilution) of the discharged effluent by jetting, buoying, and spreading the effluent; consequently, the diffusers reduce

Table 3.1 Properties of Some Models

Management Analysis	*Model*								
	QUAL2E-UNCAS	*WASP4*	*HSPF*	*SWMM*	*WQRRS*	*CE-QUAL-W2*	*MIKE:XXFLOW-3D: XXPLUM-3D*	*TRISULA-DELWAQ*	*DIVAST*
Receiving Water									
Rivers	X	X	X	X	X			X	
Reservoirs & lakes		X			X	X	X	X	X
Estuaries, coastal areas		X				X	X	X	X
Attributes									
Dynamic		X	X	X	X	X	X	X	X
Stochastic	X								
Stormwater flow			X	X			X		
Sewer system				X					
Water Quality									
Dissolved oxygen	X	X	X		X	X	X	X	
Nitrogens	X	X			X	X	X	X	
Phosphoruses	X	X			X	X	X	X	
Indicator bacteria	X	X	X		X	X	X	X	X
Suspended Solids		X	X		X	X	X	X	
Heavy metals		X					X		
Dissolved Substances	X	X	X		X	X	X	X	X
Acidity									
Temperature	X	X			X	X	X	X	
Oils, grease, PAHs							X		

the effect of the effluent on the receiving water. The diffuser design is important in any outfall (World Health, 1988) and the options available for the diffuser design are numerous. For example, a well-designed multiport diffuser located in water depths of 50 m can easily achieve initial dilutions of 80:1 and will be equally effective as a secondary treatment for biodegradable components in the wastewater in terms of receiving water concentrations.

Initial dilution is not normally included in the water quality prediction models; therefore, the initial dilution for an outfall must be predicted using models, then the prediction used as an input for the water quality models. The initial dilution is affected by the jetting velocity, direction of jets, depth of the diffuser, port diameter, number of ports and port spacing, buoyancy of the effluent, receiving water velocities (speed and direction), and stratification. Initial dilution is a complex hydrodynamic process. There are many discussions on the initial dilution hydrodynamics in the technical literature; one of the most useful is by Muellenhoff et al. (1985).

The U.S. Environmental Protection Agency (USEPA) has developed expert system models for the multiport diffuser outfalls, called CORMIX (1990 & 1993). These are available at no cost, and the user manual is particularly good in showing graphically the effects of the different variables on the initial dilution. Some model users have found that the diffuser models overpredict the initial dilution by a significant amount when the model predictions are compared to tracer studies (Sharp & Moore, 1987 & 1989). Users of the diffuser models may want to reduce the CORMIX model predictions by a factor of two to be conservative, or carry out site-specific tracer studies to quantify the initial dilution predictions of the models. It is advisable that some site-specific tracer study data be collected for any initial dilution prediction model, because initial dilution is so important in determining receiving water quality concentrations. In some instances, an existing outfall in the area can be used for the tracer studies. The data from the tracer studies can then be compared to the model predictions to determine whether a factor should be used for the model predictions. For most outfalls, and particularly for the ocean and lake outfalls, the initial dilution varies over a wide range, and it is important that the outfall prediction models be used to define this range. For some receiving water quality parameters like CBOD or indicator bacteria, mean dilutions are required, whereas other parameters like ammonia, chlorine, and metals require minimum dilutions. The initial dilution predictions have to be consistent with the objectives for the water quality parameter being modeled.

QUAL2E and QUAL2E-UNCAS (USEPA)

This one-dimensional model, designed for rivers, is a steady-state model (assumes instantaneous equilibrium). A good user's manual explains the theoretical background of the model and provides all the information necessary to run the model. The DO and nutrient kinetics in the model are complete and have been used directly in other models. The UNCAS portion of the model allows uncertainty in the input data to be incorporated directly in the model predictions. It is one of the few models that includes uncertainty in the water quality predictions and has frequently been used to identify the most important input data for the prediction and the level of precision required in these measurements. The model has been used for time-varying discharges by carrying out separate runs for a range of discharge conditions or by incorporating the variability of the discharge as an uncertainty factor in the UNCAS model. The model can be used in two-dimensional flow situations, but must be applied along the streamlines. The model temperature prediction is based on heat fluxes and is useful.

The model does not predict heavy metals, oils, grease, or PAHs. While the model can predict indicator bacteria, these predictions may not be very accurate if many of the bacterial cells are associated with organic flocs. QUAL2 does not have sub-routines for the suspended sediment dynamics; consequently, if the suspended solids concentrations are high in the receiving water, QUAL2 will have difficulty predicting some water quality parameters.

WASP4 (USEPA)

WASP was designed as a comprehensive receiving water quality model that can be used for all types of receiving waters (rivers, lakes, marine). It is a flexible model that allows users to develop their own numerical segment system and use various parts of the model. The hydrodynamics in this model are one-dimensional and time-varying; therefore, the direction of the currents must be known. The model cannot generate a two- or three-dimensional flow field, and its use is limited if the flow field is characterized by localized eddies. In other words, the model will have difficulty when applied to a receiving water with unusual topography and bathymetry, e.g., receiving waters characterized by embayments, offshore shoals and reefs, breakwaters, or headlands. The user can use another hydrodynamic model to generate the circulation pattern, then feed the output from this model into the water quality parts of WASP.

Predicting the DO, nutrient, sediment, and heavy metal kinetics is one of the great strengths of the model. The receiving water processes in the water quality parameter kinetics have been developed thoroughly in the model. (The model can also be used to predict the receiving water processes for organic substances like pesticides.) The receiving water kinetics is discussed in detail in the user's manual.

WASP is limited in its hydrodynamic capabilities, the large input data requirements to use the model, and the difficulty of quantifying the precision and accuracy of the model, which is a common problem with all complex models. Setting up the model, calibrating the model, and applying the model require extensive time for technical staff.

SWMM (USEPA), HSPF (USEPA, and MIKE 1 (Danish Hydraulic Institute)

These models all predict storm water runoff and water quality in rivers. Unlike QUAL2, these models are dynamic and predict river flow. SWMM is a sophisticated model for urban areas with a storm water sewer system capability. HSPF and MIKE 1 are similar models for rural areas and do not include sewer networks. For water quality, the SWMM model uses the WASP model for the water quality predictions, while the other models predict water quality using similar equations. These models can include the impact of point discharges other than storm water. Although the SWMM and HSPF models are sophisticated, they can be used as black box models with very limited input data. This may be a useful feature for predicting storm water runoff quantity and quality.

Except for SWMM, which uses the water quality capabilities of WASP, the models do not consider the receiving water processes in an integrated manner. The water quality aspects of the models are primarily transport, dispersion, and first-order kinetics. The models are particularly limited in the dynamics of suspended solids in terms of associated bacterial and heavy metal contaminants.

WQRRS (U.S. Army Corps of Engineers)

This is a hybrid model that meshes a river model and a reservoir (lake, estuary) model. WASP has the same capability but is deficient in the reservoir hydrodynamics.

CE-QUAL-W2 (U.S. Army Corps of Engineers)

This is a dynamic, two-dimensional version of QUAL2 specifically formulated for reservoirs, lakes, or narrow estuaries. The two dimensions are downstream and depth. This model considers all the complex processes with depth in a deep reservoir that has density stratification.

MIKE XX, TIDEFLOW-3D, XXFLOW-3D, and XXPLUM-3D (Danish Hydraulic Institute)

These models can be dynamically used in one, two, or three dimensions or as a Lagrangian spill-type model. The modeling system dynamically predicts water elevations and currents as well as the suspended solids dynamics, interactive water quality parameters that are similar to the processes in WASP. Wave-generated processes are not included explicitly in the model, but their mean impact can be simulated by adjusting the surface drag coefficients. This is one of the few models that addresses implications of **wave dynamics** in coastal water quality. These models satisfy all the water quality and receiving water requirements in this guide. It is not known whether model precision is available for the modeling system, or whether some of the models can be run in a stochastic manner, other than the XXPLUM-3D models, which are random walk models.

Like all complex models, the model setup is labor-intensive, as is the calibration process, which requires large data sets. Because the system consists of sub-models for the various components, the individual components can be set up and calibrated independently from the other sub-models. This makes it easier for the user to understand and calibrate the model.

TRISULA & DELWAQ (Delft Hydraulics)

These models predict the two-dimensional residual (tidal and wind-averaged) currents and suspended solids dynamics and all the water quality parameters of interest except for oils, grease, and PAHs as surface plumes. The water quality processes in the model are similar in structure to those of WASP, and the sediment dynamics in DELWAQ are more extensive than in WASP. The primary productivity model predicts diatom and green algal bio-masses. Predicting and using the residual currents simplifies the calibration process and is adequate for predicting algal bio-mass but not dissolved oxygen, ammonia, indicator bacteria, and metal concentrations, which have instantaneous concentration objectives or geometric averages over a number of samples. It is not known whether the models can be run in a stochastic manner or whether the models compute the prediction precision, and it may be necessary for the user to evolve a method for quantifying prediction precision.

These models can predict all the water quality parameters of interest except for the oils, greases, and PAHs as surface slicks. The setup and calibration of complex models require extensive technical staff time as well as calibration data. Predicting and using the residual currents simplify the calibration process and are adequate for predicting algal bio-mass but not dissolved oxygen, ammonia, indicator bacteria, or metal concentrations which have instantaneous concentration objectives or geometric averages over a number of samples.

DIVAST (Binnie & Partners)

This model predicts the dynamic two-dimensional currents, water elevations, and transport of a dissolved substance. The dissolved substance transport process includes first-order time kinetics. Continuously recorded currents, winds, and water levels can be fed directly into the model.

The model does not predict dissolved oxygen, nutrients, suspended sediments, heavy metals, temperature, oils and grease, or PAHs.

MODEL DATA REQUIREMENTS AND PREDICTION ISSUES

A list of the site-specific data requirements for all prediction models is presented. Because many projects lack some of the data requirements on the list, water quality prediction models can still be applied. The modeling strategies for limited site-specific data, suspect site-specific data, and non-point source loadings are discussed. The site-specific data required for a model application are discussed. Then the use of spill models is discussed.

To develop and use a water resource requires that the development be carried out in a manner that sustains the water resource for a diversity of water uses. Water resources management requires the development of water resources projects and/or management procedures to preserve and enhance water quality. Water quality prediction is the only way that different water resources management projects can be evaluated in terms of the water quality aspects; consequently, **water quality modeling is a fundamental part of all environmental assessments.**

Water quality models are designed so that they can be customized to a particular application using some site-specific data. In general, these site-specific data include physical and water quality measurements as well as some coefficients and rate constant determinations. Ideally, the user is required to provide site-specific data on the following:

1. Physical measurements of bathymetry and topography (flow cross-sections).
2. Boundary conditions (velocity, depth, flow, and concentrations).
3. Initial conditions (depth, velocity, and concentrations).
4. Discharges (location, flow, and concentration).
5. Other coefficients and rate parameters, depending on the water quality

parameter being modeled.

6. Wind and rainfall (some models).

7. One set of measured data for calibration.

8. One set of measured data for verification.

If the site-specific data above are available, the calibrated model predictions can be expected to have a precision a little larger than the sum of the precision of the measurements and measurement combinations. The precision of the measurements is defined by the quality assurance and control program.

In many water resources projects, all the site-specific measurements required are not available and some of the available data may be questionable. What are the guidelines for using water quality models in these instances?

Limited Site-Specific Data

When there are limited site-specific data, complete comprehensive complex models should not in general be used for predictions except for developing an understanding of the water quality processes in a particular receiving water or for developing a water quality monitoring program. The extensive site-specific data requirements for comprehensive complex models would render any determination of the precision of prediction meaningless based on the use of literature values. Simpler models or simplified comprehensive models should be used. The following discussion is intended to assist in identifying the appropriate model simplifications or the preferred model for a particular application. In this process, the difficulty in measuring some of the site-specific data discussed previously is considered.

Some site-specific measurements are necessary for any model application. The basic requirements are 1, 2, and 4 above. Cross-sectional data are required at the upstream and downstream boundary locations and at a minimum of three locations in between. These measurements should be for a common flow and/or water depth conditions. If not, these measurements should be adjusted to the same flow conditions using standard hydraulic techniques. Water depths, flow, and concentration data must be available at the upstream and downstream boundaries and at some intermediate cross-sections.

Again, these measurements should be for the same conditions of flow and/or water depth as the cross-sectional data and if photosynthesis and respiration are factors at the same time of day.

To determine if photosynthesis and respiration are factors, dissolved oxygen and temperature (and salinity for marine waters) measurements should be made over a 30-hour period during the aquatic plant growth season. If this is not possible, measure DO and temperature at several locations early in the morning and at mid-day. If the percentage change in DO percentage saturation in these measurements is significant, photosynthesis and respiration are factors in the DO, and nutrient kinetics in the receiving water must be considered.

The location, flow, and concentrations in the discharges must be known or estimated. The water quality parameters of interest will be identified in the preliminary water quality measurements or other data. Select the simplest appropriate model from Table 2.2 (remember that dynamic models can be run as steady-state models). Most models have default options and/or provide a range of values for bottom roughness, eddy diffusivity, dispersion coefficients, and the other required model coefficients and rate parameters. Because a calibration data set is not available, it is not possible to determine the values of the coefficients and rate constants required to apply the model. These values will have to be selected from values provided in the manual or some other source. To estimate the impact of this selection on the predictions, it will be necessary to use the model several times with different values for the coefficients (sensitivity analysis). It is suggested that the model be used with the coefficients/rate parameters in the range of (mean ± (0.17 x range)) to quantify the precision of the predictions. The model in this form may be very useful in providing guidance for designing an appropriate monitoring program for the model.

If a partial site-specific data set is available, the missing data can be selected from the range of values in a manner similar to that discussed above.

Another approach is to use a stochastic type model. In these models, the range or mean value and distribution (normal, lognormal, etc.) of the user input data can be provided, and the models can be used stochastically. The predictions from these models will include mean value, range, and distribution; consequently, the precision of the predictions is included in the output. The stochastic form of the models is the best for limited data sets (Zielinski, 1988).

Suspect Data

In many instances, the data available in a particular project have been collected at different times, different locations, and by different agencies using different laboratories. Without some kind of quality assurance and control program in place, it is difficult to assess the validity of any data. Some simple water quality checks can be used (for example: total organic nitrogen includes ammonia; fecal coliforms include *E. coli;* bacterial density measurement based on a single sample is very questionable [coefficient of variation about 0.3 log]; if suspended solids concentrations are high, single samples for total heavy metal concentrations are questionable; laboratory analysis of dissolved oxygen concentrations, pH, and turbidity are questionable, etc.) but these are limited. It is important to **remember when reviewing data that some or all of the measured data may be questionable. Equally, the model used for the predictions may be incomplete and missing some of the important receiving water processes; consequently, the model can produce erroneous predictions.** Similarly, the precision and accuracy of water quality data must be estimated. Either one can be faulty. If there is a difference of more than one standard deviation between the measured and the predicted data, one of the two is probably questionable. Identifying which one is questionable requires checking the predictions of the model with other measured data sets.

Non-Point (Runoff and Groundwater) Sources

Some of the models in the Appendix predict runoff and groundwater flows in a sophisticated scientific manner. The export of suspended solids in the runoff is also predicted in the models; however, the concentrations of water quality contaminants in the runoff and groundwater are based on statistical results for different types of land use. It has been shown in numerous studies that **about 15 different storm events have to be measured before the statistical runoff characteristics of a catchment can be defined**; therefore, defining non-point source loadings for a particular catchment can be costly. Model users must account for non-point sources in the prediction process, but must be aware that estimates of such loadings may have a large error (≈40 percent for nitrogen, ≈60 percent for phosphorus and heavy metals). Water quality predictions for periods when runoff is not a significant factor will be much more reliable.

Designing a Water Quality Monitoring Program

Once a model has been selected for a particular application, it can be used to determine the most important parameters that should be measured to improve the reliability of the prediction. Using the procedures discussed previously for the limited site-specific data, a prediction model will have been developed and the precision of the model predictions will have been defined. The next step is to vary the other input variables like boundary conditions, discharge loadings, cross-sections, etc., and determine the sensitivity of the model predictions to these variations. This process will identify the most important monitoring requirements for the application of the model. As discussed previously, some of the water quality monitoring requirements are very tedious and costly to carry out, and it is important to determine where these measurements are necessary for the prediction of the water quality. For example: Is it necessary to predict photosynthesis and respiration? Is it necessary to measure sediment oxygen demand and surface re-aeration? Is it necessary to measure all the discharges?

QUALITY ASSURANCE AND QUALITY CONTROL

Monitoring Data

Water quality monitoring requires either a field measurement in situ (e.g., temperature, pH, dissolved oxygen, turbidity, depth, velocity) or the collection of water samples that are analyzed in the laboratory for concentrations of various water quality parameters. Standard laboratory methods have been developed for most water quality parameters; these are published by various regulating agencies. In some instances, field sampling and field handling procedures of the samples between the sampling location and the laboratory are also specified by some regulating agencies. These procedures reduce variability in the sample collection, transport, and laboratory analysis processes carried out by different personnel in different receiving waters. To quantify the variability associated with sample collection, transport and laboratory analysis, most regulating agencies have established quality assurance and quality control (QA/QC) procedures consisting of the following (see for example USEPA, 1990):

- ≥10 percent field replicates* and ≥8 replicate samples(for micro-organisms, all samples should consist of at least triplicates;some regulating agencies specify the geometric mean of 5 to 10 samples for micro-organisms, in which case it may be necessary to collect 5 to 10 replicate samples);
- ≥10 percent laboratory splits* and ≥8 replicate samples;
- ≥5 percent blanks* for both field and laboratory blanks;
- replicate calibration against standards or spikes and/or interlaboratory sample; and replicate determination of detection levels if not defined in the standard method procedures.

* For conventional water quality parameters, ≥5 percent has been found adequate for field and laboratory splits and ≥2 percent for blanks.

QA/QC control procedures quantify the precision and accuracy of the water quality data for each measured water quality parameter. Every sampling survey must have quality control data because the precision and accuracy of the water quality data can be different for each survey. The detection limit for most water quality parameters is about 5 times the highest blank concentration; for volatile water quality parameters, it is about 10 times. Blank concentrations should never be subtracted from measurements. The standard deviation of the laboratory splits defines the precision of the analytical technique, and the standard deviation of the field replicates defines the precision of the field sampling, sample handling plus the precision of the analytical technique. Precision is normally defined as a coefficient of variation (%) = (standard deviation x 100)/ mean concentration. The accuracy of the analytical method is defined by the calibration against standards and is defined as the standard deviation of at least eight calibrations. The accuracy of various standard methods is normally provided in the procedures description. It is acceptable to use these published accuracy data. For many surveys, quality control data are also required for positioning, timing, and depth following the guidelines for field replicates. A general guide of ≥10 percent for field replicates should be followed.

The precision and accuracy of water quality data must be quantified using quality assurance and quality control procedures. If these data are not available, literature values can be used and the literature referenced.

Model Prediction

Water quality modeling predictions also require quality control, but specifying quality control procedures for modeling is much more complex than for water quality monitoring (Barnwell & Krenkel, 1982; Simons, 1985; Benarie, 1987; Ellis et al., 1980; Sharp & Moore, 1987). Both the characteristics of the model and its application affect the selection of the quality control procedures.

All models have coefficients or rate constants or factors what are required for the model to generate predictions—the more complex the model, the greater the number of the coefficients. Ideally, the coefficients should be site-specific and determined from local field data in the model calibration and verification processes. The inherent precision of the water quality data must be considered in the calibration and verification process. In complex numerical models that are time-variable and in one-, two-, or three-dimensional space, the coefficients must be defined at each solution point and time step. The calibration and verification processes in complex models are laborious trial-and-error procedures (Rasmussen & Badr, 1979). Sometimes it is possible to simplify the models by carrying out a sensitivity analysis of the model input requirements to determine the most important parameters in the model. In the sensitivity analysis, the model input parameters are varied over a small range to determine the effect of these changes on the predictions. The results of **the sensitivity analysis can be used to determine the parameter measurement requirements** and/or the feasible model simplifications or may indicate that the selected model is not suitable.

If field data are not available, many models provide default options or typical values that can be used in the models. QA/QC procedures are needed for these coefficients so that the precision in the predictions associated with the selection of the coefficient values can be quantified. Some models can be used stochastically, which includes the variability of the coefficients in the prediction (see Dewey, 1984; QUAL2E-UNCAS in Appendix A or QUAL2, 1987; Zielinski, 1988).

Many of the model predictions require the numerical solution of individual or coupled partial differential equations normally on a spatially defined grid. It is necessary to select the grid or element size and the time step boundary conditions to obtain solutions to the equations. These variables must be selected in such a way that the mathematical solutions are stable and converge rapidly; however, the selection process may affect the precision of the predictions. In general, longer time steps have less numerical dispersion. The precision for these aspects (spatial and time scales) should be quantified in the procedures.

In many instances, water quality prediction models are used to compare the effects of different water management scenarios, typically capital works projects. Predictions for these applications are normally presented as percentage improvement or degradation between one scenario and another. While the difference between two model predictions is more precise than a single model prediction process, there still is a need to quantify the precision of the differences. For example is a 5 or 10 percent difference in the predictions greater than the precision for the prediction process?

Defining the precision for the prediction process is also important in determining the level of the model prediction and the type of model that is the most appropriate for a particular project. For example, the precision of a three-dimensional model may be too large for a particular application. One way to improve precision is to simplify the model and/or reduce the number of dimensions to two or one.

Unlike the procedures for water quality monitoring, which can be defined generically for different water quality parameters, **quality control procedures must be developed for each model application.** The objective of quality control in the modeling exercise is to define the precision of the model predictions for a particular application (Rasmussen and Badr, 1979). Ideally, the precision should be evaluated for the model components like hydrodynamics, mass balance, receiving water process, and sediment dynamics, separately if possible, because the magnitude of the precision can be different for each of the components. This information is useful in providing direction for future monitoring and modeling efforts. If one of the modeling components has a precision much greater in magnitude than the other components, the precision in this component can be improved by increasing the monitoring effort for this component or by simplifying the model. In evaluating the component precision, it is necessary to account for the dependence of one model component on another component; e.g., the mass balance component uses the output from the hydrodynamic component and, consequently, includes the precision for this component. For the selection of the appropriate model and the level to be used in that model, the following procedure could be used:

- List the available site-specific data.
- Summarize the historical information on water quality problems or degradations. This information should include both water quality measurements (quantitative) and qualitative data.
- Visit the site to confirm historical information on water quality and to note any special factors in the receiving water that relate to water quality modeling. These factors could include observations of visible surface slicks, receiving water color or turbidity, aquatic plant growths, backwater areas, recreational swimming or fishing, visible bottom sediments, private domestic sewage discharges, irrigation withdrawals, livestock watering or crossing, etc.
- Interpret the available site-specific data, historical information, and site visit information.
- Quantify the precision and accuracy of the available data. If quality control data are not available, use literature values for the method used to measure the data. If only laboratory analysis precisions are available, use 1.2 X (laboratory precisions) for the precision for field plus laboratory analysis.
- Select water quality models that are suitable for the project (i.e., models that satisfy the water quality prediction objectives).
- List the input requirements for each model candidate.
- List the model input requirements for which there are no site-specific data available. Based on the interpretation of the available data, histor-

ical information, site visit, and list of the required model inputs that are not available locally, simplify the candidate models and then select the most appropriate model(s). A site visit is extremely important in the model selection process.

- Set up the necessary topographical grids for the selected model(s).
- For the missing input requirements for the selected model(s), determine a range for each input using published values or values from related projects. Determine the range of values for the available site-specific data. Now predict the water quality concentration with the model (C_A) using one-third of the range for all input parameters. Then predict the water quality concentration with the model (C_B) using two-thirds of the range for all input parameters. An estimate of the model precision expressed as a percentage = $(C_A+C_B)/((C_A+C_B)/2)$. Alternatively, it can be assumed that the coefficient of variation is on the average 15 percent for all input requirements. If six to eight separate model predictions are made randomly selecting values within ± coefficient of variation, the standard deviation of the predictions is a good estimate of the model precision (Dewey, 1984).
- If the model precision is within 50 percent of the site-specific measured data, the selected model probably has the appropriate sensitivity for project use.

Other more rigorous methods to determine the model prediction precision are preferred for any particular model and its application. A method can be used at the discretion of the model user. For the non-linear numerical models, some method based on chaos or sensitivity analyses may be appropriate. It is not possible to quantify the precision of a model prediction with a single prediction verification for deterministic or numerical models. One of the best methods for quantifying precision is to use the stochastic or Monte Carlo formulation of the model(s). **In some instances, it may be appropriate to use a stochastic model as an additional instrument to provide an estimate of the precision if stochastic forms of the models are not available.**

Any selected model may not include some of the important processes in the receiving water; therefore, the prediction will be imprecise. For example, a model may not include the impact of the resuspension of bottom sediments on water quality, and this may be

Typical Sensitivities for Specific Water Quality Parameters

Parameter	Sensitivity
Temperature	2–3%
BOD	10–20%
Dissolved oxygen	5–10%
Nitrogen	15–30%
Phosphorus	15–40%
Algal	10–25%
Bacteria	0.2–0.35 log
Conservative	5–10%
Heavy metals	25–50%
Cohesive sediments and flows	50–100%

the major loading source in some receiving waters. Or a model may not include atmospheric loadings. These model formulation discrepancies will probably not become apparent until the model predictions are shown to be consistently different from measurements. This is one of the reasons that multiple verifications are required to be sure that the model selected contains all the most important variables.

In some instances, the precision of the prediction for the water quality parameter of interest may be very large (for example, within one order of magnitude) and therefore not very useful for water quality management; e.g., indicator bacteria, phosphorus, heavy metals. In these instances, the model user can carry out numerous predictions to statistically improve the precision, or simplify the model, or use a surrogate parameter. The surrogate can be used for predicting transport, dispersion, and settling. Then, the surrogate concentrations can be related to the concentrations of the parameter of interest through studies. Or the surrogate can be used to predict the transport, dispersion, and settling to which have been added the other processes relative to the water quality parameter of interest, like mortality rate for indicator bacteria. Suspended solids concentrations have been used extensively as a surrogate parameter for storm water quality. In sanitary wastewaters, suspended solids concentrations are about 200 mg/L. Dissolved solids concentrations as measured by specific conductivity have also been used as a surrogate parameter for wastewaters. The conductivity of sanitary wastewaters is over 1,000 umhos/cm.

The precision and accuracy of water quality data must be quantified using generally accepted quality assurance and quality control procedures for water quality modeling. The precision and accuracy for the model predictions must be developed for each model application. Stochastic forms of the prediction models are useful in quantifying the model prediction precision and accuracy.

Spill Models

Spill models were discussed previously. Normally, these models can be used with little site-specific data; however, the same Lagrangian spill models can also be used to predict a plume, which can be on the water surface or at depth. In the Manila Second Sewerage Project, for example, site-specific data were used to generate the currents at the ocean disposal site, as well as background measurements of the ocean concentrations in the vicinity of the proposed dumping sites. The resulting ocean plumes were predicted for different conditions and different locations so that the optimum disposal site could be selected. The spill models are required for short-period discharges or releases of substances that can degrade the receiving water quality. In this instance, a quantity of tracer dye is instantaneously released, and the tracer concentrations are measured in the tracer plume at various times thereafter. Using a spill-type model, the dispersion coefficients can be determined from the tracer plume measurements.

CHAPTER 4

Case Studies of Models Applied to World Bank Projects

This chapter presents seven diverse case studies from China, India, and the Philippines. The case studies were taken from projects implemented with funding by the World Bank. In each case, the modeling work was necessary in order to allocate scarce financial resources so as to maximize environmental benefits.

DETAILED HYDRODYNAMIC AND WATER QUALITY MODELING STUDY, 1998, CHONGQING, CHINA

The municipality of Chongqing has a population of about 30 million in an area of 82,000 km^2 along the Yangtze River and Chang Jialing River (the third longest river in the world, with a mean annual flow at Chongqing of about 100,000 m^3/s) upstream from the Three Gorges Dam. The City of Chongqing is located at the confluence of the Yangtze and its major tributary the Jialing.

Water supplies for the City of Chongqing are derived primarily from the Jialing and, to a lesser degree, from the Yangtze. These rivers also receive discharges, largely untreated, of municipal and industrial wastewater from the city. These discharges are poorly dispersed in the river water, and some are upstream from water supply intakes. Thus, despite the very large flow and assimilative capacity of the rivers, the quality of raw water supply, particularly from the Jialing, is often poor.

The construction of the Three Gorges Dam will increase the water depth of the Yangtze by up to 16m and significantly reduce the velocity of the river to less than low-flow conditions. Major investments were therefore required for wastewater management infrastructure in the City of Chongqing. The magnitude of the associated costs necessitated a phased approach to the provision of sewerage system improvements and wastewater treatment plants.

Secondary treatment of all wastewater would meet water quality objectives but at a high cost, in part because of the high cost of land in the urban area. The water quality problems of the Jialing could, however, be readily and economically solved by improved collection of wastewater and discharge into the Yangtze downstream of the confluence. This would effect substantial environmental improvement within the urban area of Chongqing at a cost significantly lower than that of full treatment of all wastewater discharges.

The Yangtze, however, provides water supplies for a large population in its catchment downstream of the Jialing confluence, including the important cities of Fuling and Wanzhou. It was therefore necessary to determine how such an option would affect water quality in the Yangtze in the immediate term and, if determined to be beneficial, the time scale within which wastewater treatment would need to be introduced.

Summary of the Project

These issues were investigated by means of mathematical water quality modeling for the Yangtze and Jialing, with two objectives:

1. to evaluate the effects of wastewater discharge (at various levels of treatment) from Chongqing on water quality in the Yangtze in a far-field context to determine the level of compliance with water quality objectives for the Yangtze and the need for wastewater treatment; and

2. to use near-field modeling of the effects of wastewater discharges in the Yangtze to optimize the discharge arrangements so as to minimize the negative effects on water quality by efficient mixing.

The modeling was undertaken by Chongqing Environmental Protection Bureau and the Danish Hydraulic Institute (DHI), with financial assistance from the Government of Denmark. The DHI MIKE system of models was used in the one-dimensional form for the far-field and two-dimensional form for the near-field modeling.

The one-dimensional model MIKE 11 was set up for the Yangtze and Jialing Rivers. In this application, MIKE 11 consisted of four components: NAM (a precipitation runoff model), HD (a hydrodynamic model), AD (an advection and dispersion model), and WQ (a water quality model). The governing equations in each model are presented in this guide. These equations are a set of partial differential equations which are numerically solved at all the grid points in a topographical network established to represent the river using a time- and space-centered finite difference scheme to minimize the numerical dispersion in the solution. The spatial and time steps used in the solution are selected to ensure that the solution is stable and rapidly converges. The required inputs for each model are presented in the report.

The models were then calibrated on the basis of data available for 1986-87 and verified if possible against the 1993-94 data. Calibration is the process of adjusting coefficients, initial conditions, and boundary conditions so that the predictions match the calibration data set. For the hydrodynamic model including NAM measured and predicted water levels, river flows and accumulated flows were matched. The calibrated parameters are presented in tables, and the calibrations shown graphically in this guide. The AD component was calibrated by adjusting dispersion coefficients and initial and boundary conditions for dissolved substances. The calibration process for dynamic finite difference schemes is a laborious trial-and-error process.

The processes in the water quality component of the model for dissolved oxygen include biochemical oxygen demand loadings; rates of BOD degradation, re-aeration, photosynthesis, and sediment oxygen demand; nitrogen loadings; rates of nitrification, denitrification, sedimentation, and resuspension; and the uptake by algae and plants. The water quality model predicts temperature and concentrations of DO, BOD, ammonia- N, nitrate-N, phosphorus (as P), and fecal coliform. The most important calibrated parameters are presented in tables, and the calibrations presented graphically (Figure 4.1). Establishing the point and non-point discharges to use in the model was not a simple task, because flow and concentration data were not available for all discharges to the river. Figure 4.2 shows the discharges used for the modeling for the Jialing River.

Figure 4.1 Simulated Concentrations Along Jialing River 1987

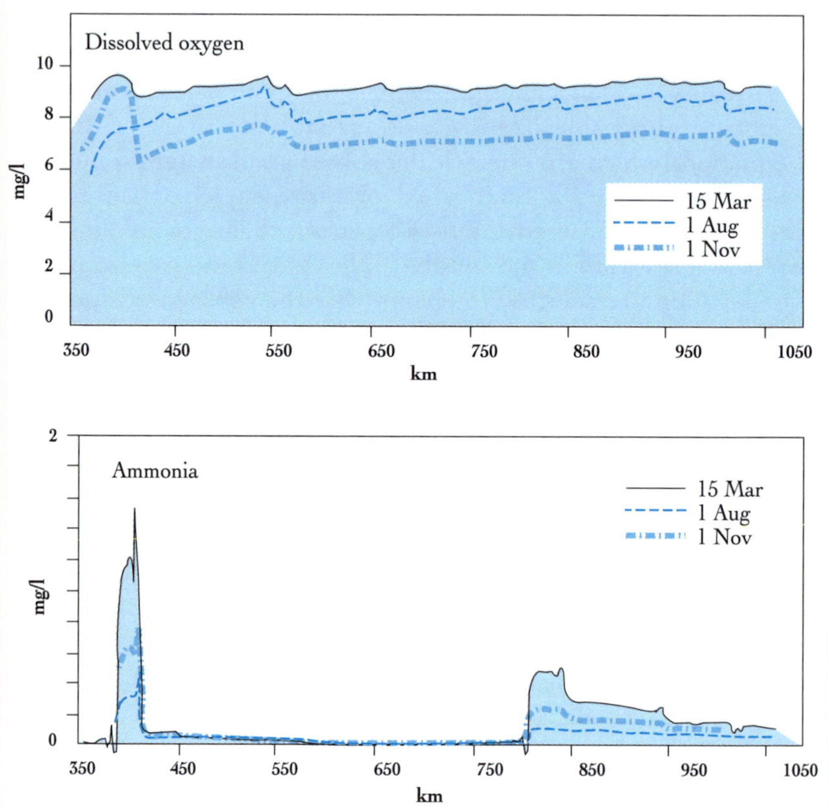

The calibrated model was then verified using the 1993-94 data set. It was found that the simulated water levels and measured level were in good agreement during the dry season. There were some differences at some locations during the wet season. However, because the accumulated flows during the dry season were within 5 percent of the measured data, the calibrated model was considered appropriate for evaluating the water resources alternatives. There were more differences between the simulated and measured water levels, with the simulated as much as 1.5 m below measured. The difference in water levels and river cross-sections was considered to be insignificant because the simulated water levels were less than those meas-

Figure 4.2 Schematic of Source Loadings

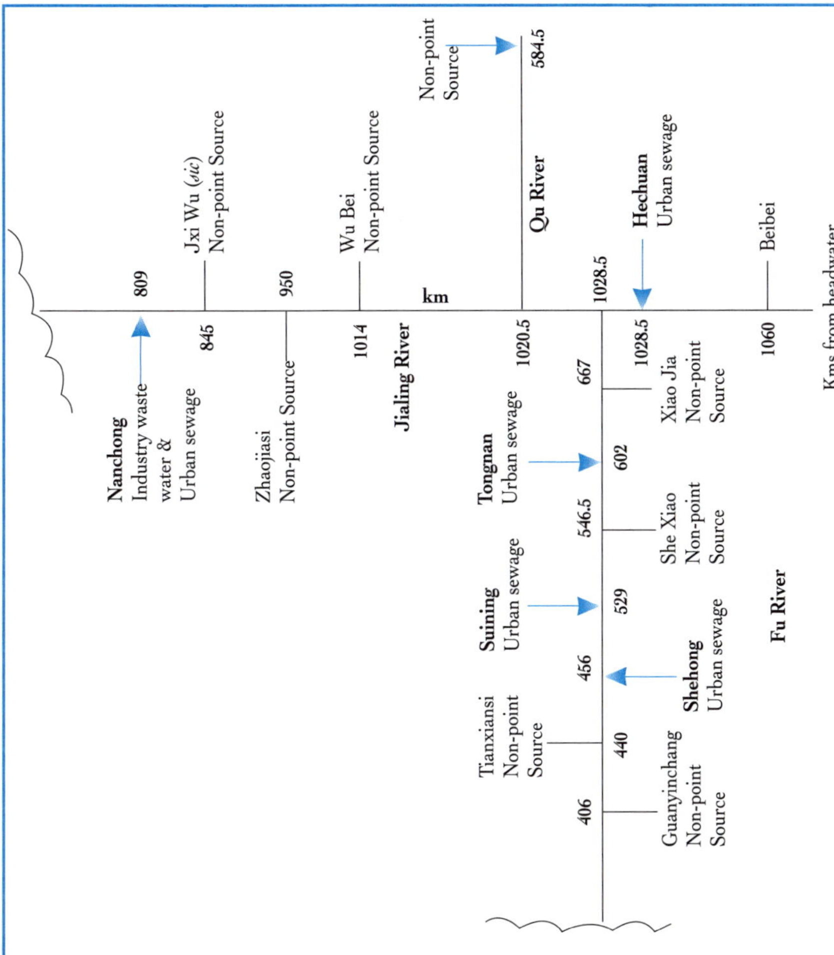

Note: Axis numbers indicate distance in kilometers from headwater

Figure 4.3 Scenario 2 with Treatment Plants

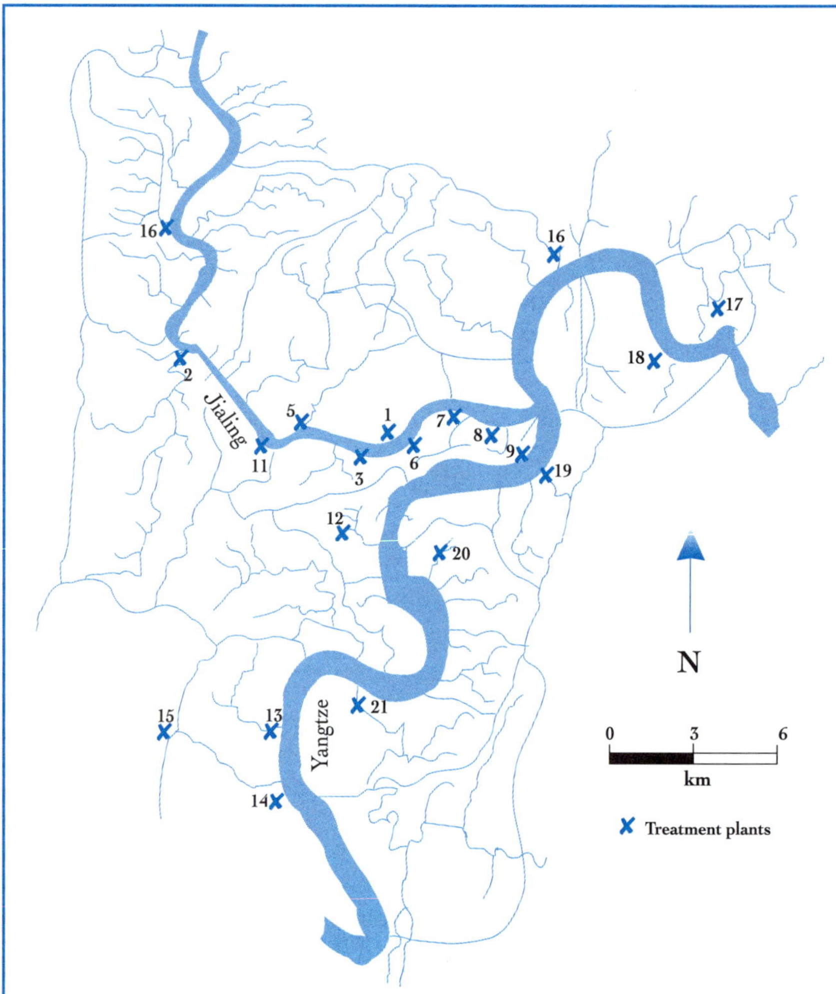

ured, using the model in the calibrated form will be a conservative application.

The calibrated models were used to evaluate the five scenarios involving different combinations of wastewater collection and treatment. Figures 4.3 and 4.4 show two of the scenarios (Nos. 2 and 5). Scenario 5 consists of interceptors along the Jialing and Yangtze

Figure 4.4 Scenario 3 with Interceptor Along Jialing River

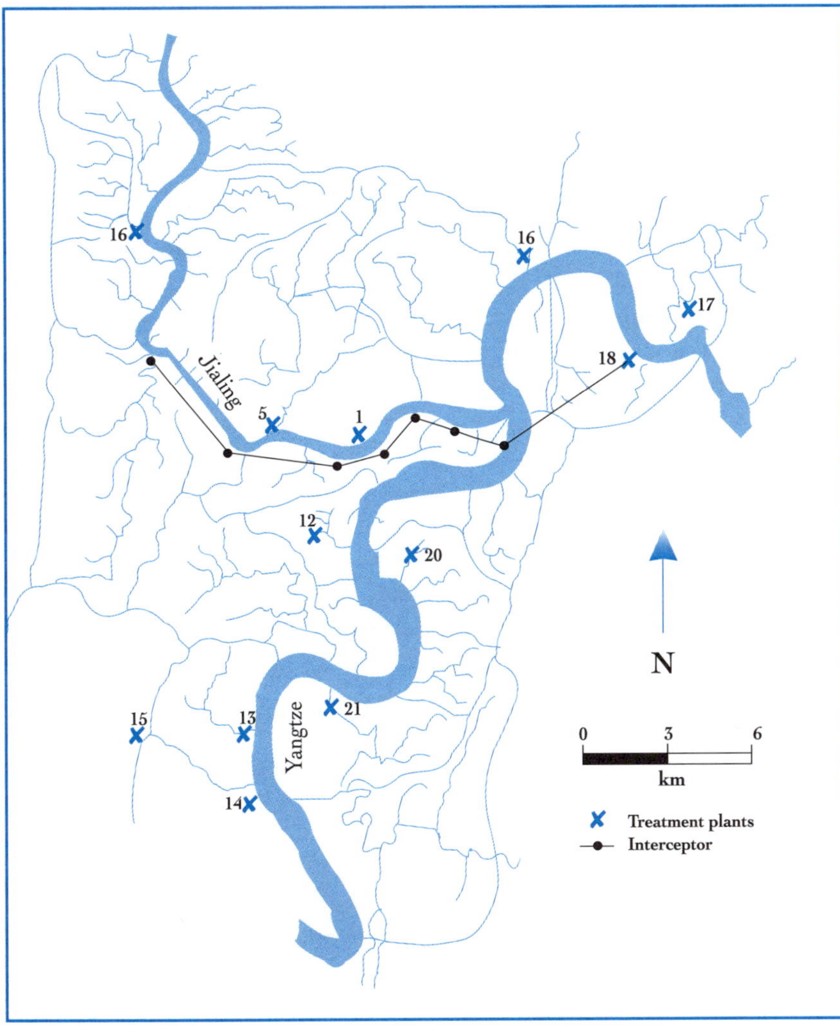

Rivers and wastewater discharge downstream from the Chongqing urban area. Scenario 2 consists of treatment of wastewater in a large number of treatment plants, with the local discharge of treated wastewater into the Jialing and Yangtze.

The far-field (1-D) modeling results showed that it would be possible to meet the desired water quality objectives for the Yangtze in terms of DO, BOD, and ammonia until at least 2010, with only preliminary treatment of Chongqing wastewater discharged to the Yangtze downstream from the confluence with the Jialing. Subsequently, wastewater treatment would need to be introduced, in particular to control nutrient loads.

The modeling enabled development of a phased program of wastewater infrastructure for the city based initially on wastewater interception and conveyance by a system of major tunnels for discharge to the Yangtze. An overall treatment strategy was developed for Chongqing to sustain and protect water quality in the Yangtze and to prepare for the probable requirements for control of nutrient loads to the Yangtze in the light of possible eutrophication of the Three Gorges Reservoir.

The local water quality in the river reaches through the Chongqing area was evaluated with a two-dimensional model, MIKE 21. MIKE 21 is similar in structure to MIKE 11, with HD, AD, and WQ components. The equations used in the model are presented in this guide. The model was run for five-day periods and was calibrated using 1987 data and verified using 1994 data. The velocity vectors in the river were simulated. Figure 4.5 shows the simulated maximum ammonia concentrations.

The 2-D near-field modeling identified the optimum discharge location and diffuser configuration for the discharge of the wastewater into the Yangtze to minimize the mixing zone required. It particularly highlighted the importance of discharge via a diffuser system into the deep river channel.

Costs

The project costs were about US$400 million. Modeling costs of US$500,000 were provided by the governments of Chongqing and Denmark. This is equivalent to only 0.1 percent of the overall project cost, but importantly, the modeling confirmed the applicability of the proposed phased program, resulting in savings on the order of hundreds of millions of dollars.

Figure 4.5 Simulated Maximum Concentrations of Ammonia in January 1987

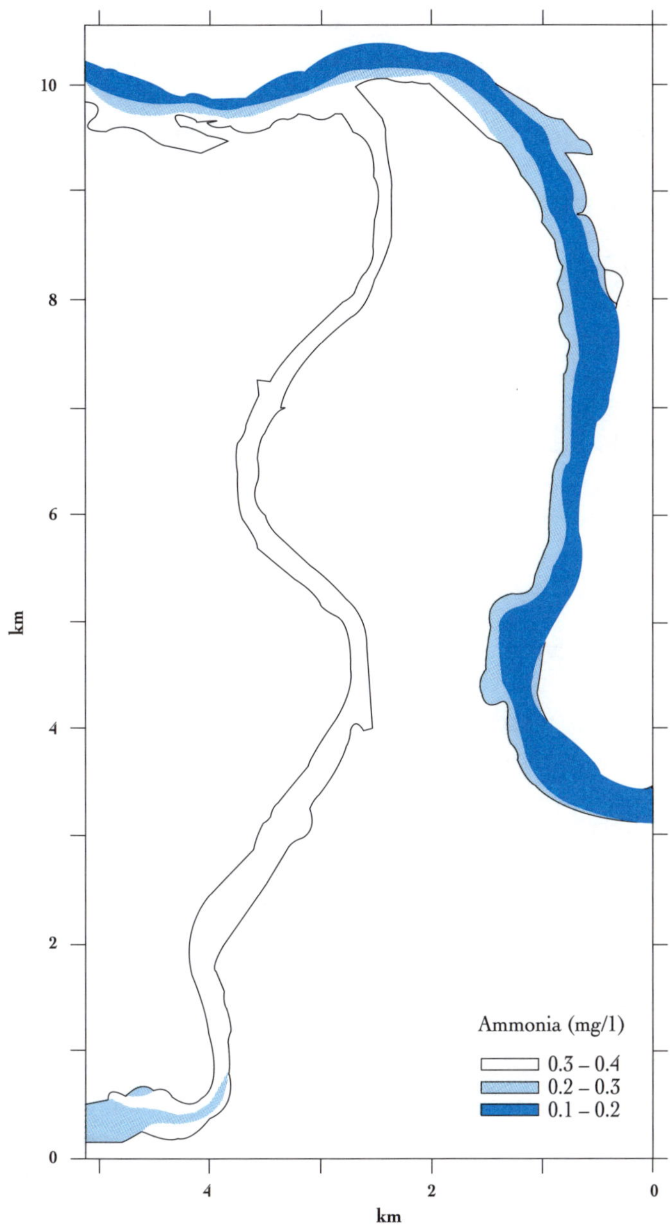

OCEANOGRAPHIC AND WATER QUALITY MODELING STUDIES AT MUMBAI, INDIA, 1997

The island city of Mumbai, India, generates about 2 million Mm^3/d of sewage, most of which is discharged untreated into the Arabian Sea. As part of the Bombay Sewage Disposal Project, design and environmental assessment (EA), some limited modeling work was carried out during 1990-95 to assess the proposed project's effects on the water quality of the coastal areas and tidal creeks. The EA recommended more extensive studies, including development of mathematical simulation models to evaluate treatment and outfall options for the next stage of sewerage investments in Mumbai. Based on these recommendations, a comprehensive coastal water quality study was commissioned in 1995. It developed an integrated hydrodynamic and water quality model to simulate fecal coliform densities for 35 km of shoreline and extending offshore for 15 km; to collect bathymetry, tidal, and current data appropriate to calibrate the prediction models for both monsoon and non-monsoon periods; and to carry out tracer and indicator bacterial decay measurements.

Water Quality Modeling Requirements

The requirements were, using collected physical and water quality data in the area of concern, to develop and calibrate a water quality model suitable for assessing the effects of different levels of treatment and different outfall lengths and locations on the coastal water quality. The assessment process was required to identify the most cost-effective pollution abatement option for enhancing the coastal water quality.

Summary of the Project

Binnie and Partners, Ltd., of the UK had carried out the original limited assessment studies, and the National Environmental Engineering Research Institute (NEERI) of India was contracted by the Municipal Corporation of Greater Mumbai to carry out the detailed water quality modeling, using Binnie and Partners' model, DIVAST (Depth Integrated Velocity And Solute Transport), through a subcontract. This is a dynamic two-dimensional finite difference model which assumes that the circulation in the area of interest is primarily horizontal (parallel to the water surface) and not stratified with depth. A basic grid size of about 200m by 200m was used for the modeling area, which extended 35 km along the shore and 14-17 km offshore but varied in some areas to reduce the computational time in the model. Typically, the numerical model was run for 48 hours of real time. Data from recording tide gages (3 in the model area) and recording current meters (10 in the model area), as well as measured bathymetry and topography, were used as inputs to the model (see Figure 4.6). The open model boundary to the north used recorded water levels, and the open boundary to the south recorded currents. Wind data were available from two shore-based locations.

The bottom roughness coefficients, surface wind drag coefficients, eddy coefficients, and model time steps were adjusted to match simulated and measured currents and water levels in the model area (see Figure 4.7). To calibrate the solute model components, tracer plume data were available for rising and falling tides at three locations using Rhodamine B and at one location using the radioisotope ^{85}Br (half-life 36 hours) in an ammonia bromide solution. The decay rates for the indicator bacteria were measured in the laboratory twice for simulated receiving water conditions. The solute coefficients, like dispersion coefficients, were adjusted in the calibration process so that the model-simulated predictions corresponded to the measurements. Once the models were calibrated, they were used to simulate monsoon and non-monsoon conditions (see Figure 4.8).

The model was used to assess the impact of primary and secondary treated effluent discharged at various distances offshore (3-88 km) for the spring and neap tidal conditions and both the maximum and minimum initial dilution conditions at the outfall. Figure 4.9 shows the variation of fecal coliform densities for the primary treatment scenario.

The model simulations showed that the wastewater plumes were greatest in extent during the non-monsoon periods. At two locations, extending the outfalls farther offshore did not significantly reduce the size of the effluent plumes owing to local circulation patterns. The simulations also showed that secondary treatment would not significantly improve the nearshore water quality at Worli and Bandara.

Figure 4.6 Current Meter and Tide Gauge Locations and Model Area

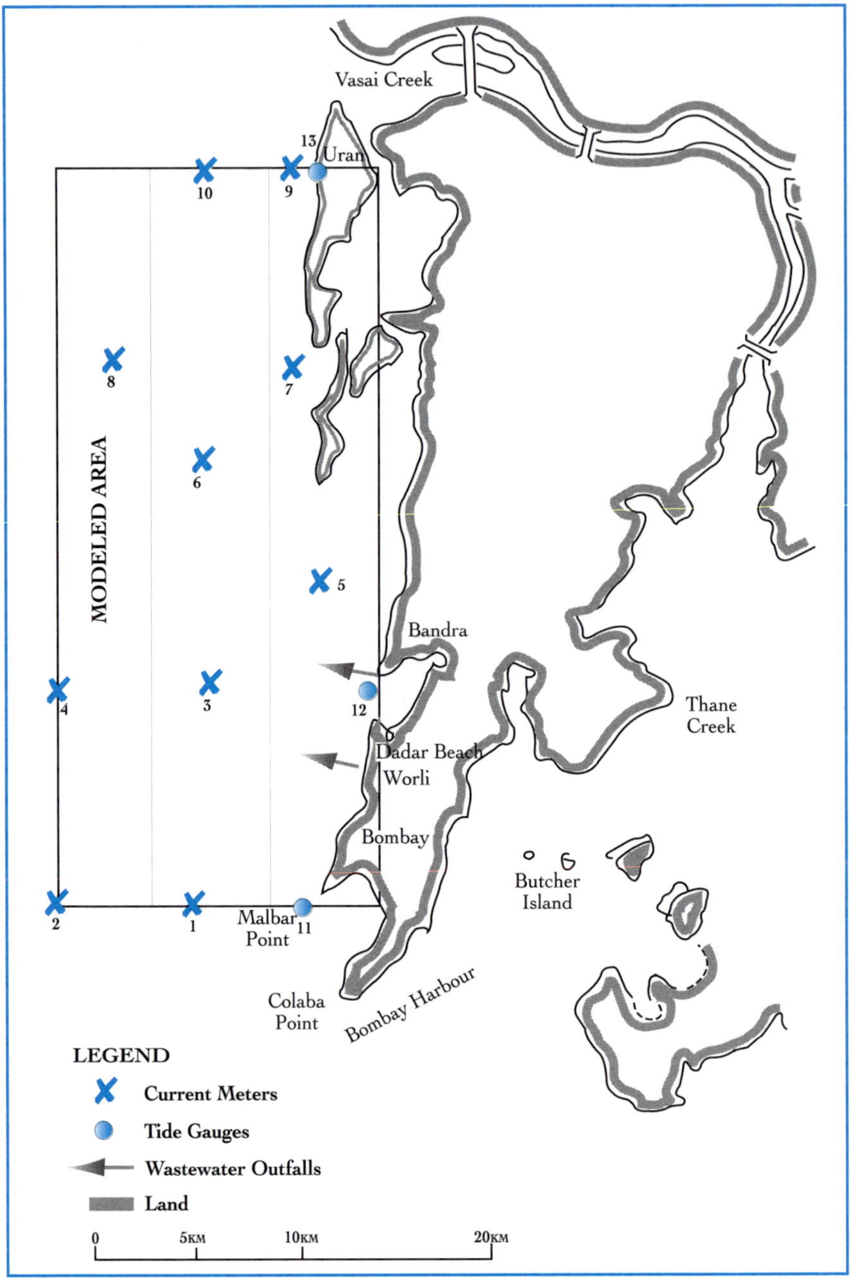

Figure 4.7 Calibration Curve for Velocity and Direction Spring Tidal Condition

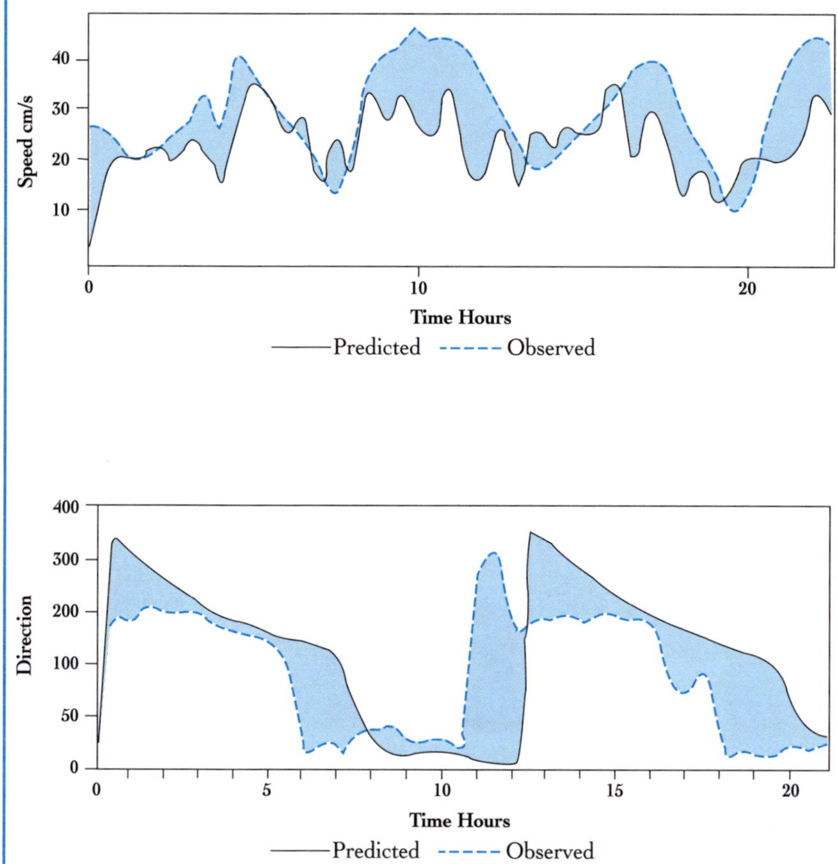

Figure 4.8 Fecal Coliform Densities at 3 and 8 Km for Primary Treatment

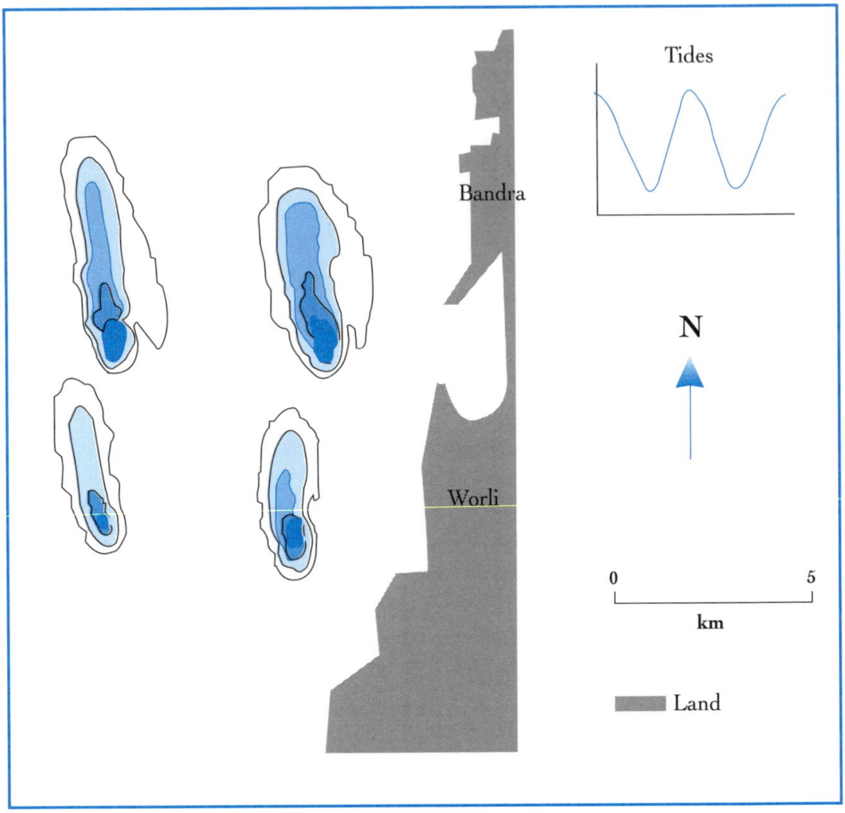

Costs

For 1996, the estimated costs for Worli were US$60 million for primary treatment and US$212 million for secondary treatment; at Bandara, the costs wereUS$97 million and US$286 million, respectively. The outfall costs wereUS$8.7 million per 500 meters. The complete coastal water quality modeling studies cost were about US$880,000. Financing came from a PHRD (Japan) grant administered and managed by the World Bank, with contributions from the City of Mumbai and from NEERI, the study consultants.

Figure 4.9 Hourly Variation in Fecal Coliforms Near 3 Km Worli Outfall

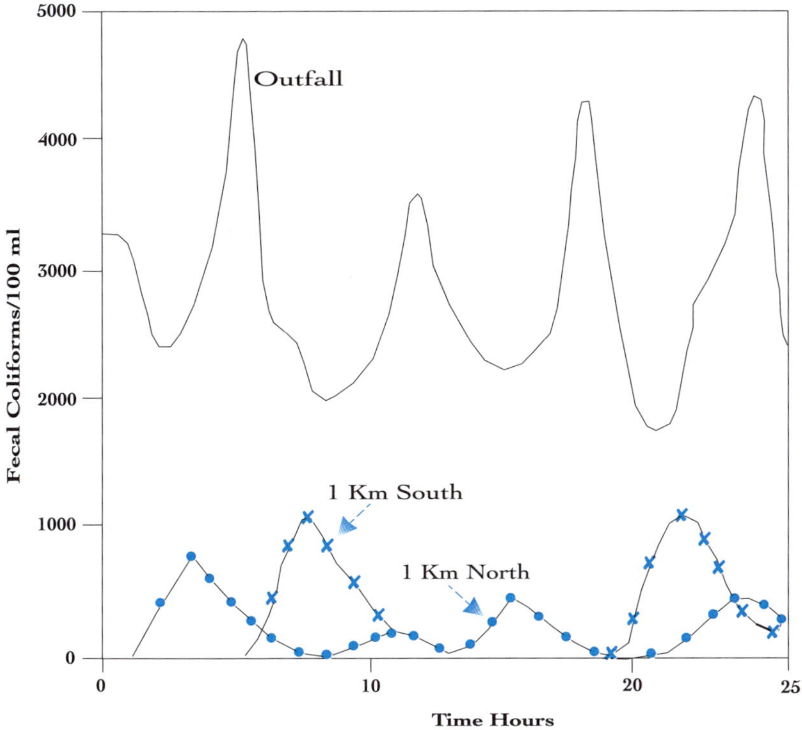

HANGZHOU BAY ENVIRONMENTAL STUDY, 1993–1996

The goal of the study was to assess the environmental impact of multiple activities occurring in and around Hangzhou Bay, eastern China, and to determine the critical balance that must be maintained among local industrial growth and economic expansion, intensive inland agricultural activity, and preservation of a viable fishery of national importance. Four key tasks were identified:

1. Define the baseline environmental conditions.

2. Estimate the environmental impact of ongoing development around the bay.
3. Develop strategic plans and policy statements.
4. Consider the planning options over a 20-year period.

Water quality models were extensively used to assess the various non-intervention and intervention options.

Water Quality Modeling Requirements

A hydrodynamic numerical model was required to simulate the circulation patterns in the bay for different tidal conditions, river flows, and wind climates. Superimposed on the hydrodynamic model, a water quality model was required to predict sediment transport/deposition/resuspension, nutrient and phytoplankton dynamics, dissolved oxygen/biological oxygen demand kinetics, and heavy metal dynamics. The calibrated hydrodynamic and water quality models were then used to assess the merits of various intervention options.

Summary of the Project

The modeling was carried out by Delft Hydraulics and Mott MacDonald international consultants, using Delft's TRISULA hydrodynamic model, a two-dimensional finite element model using a curvilinear grid (see Figure 4.10). The tidal flow was calibrated first, followed by a salinity calibration. The methods for generating model input data at the open sea boundaries of the model to the east and south (no site-specific data were available), determining the best values of bottom roughness (no site-specific data were available), and determining the best eddy viscosities are discussed in this guide. Open boundary data and bottom roughnesses were adjusted to match measured water level data, and the eddy viscosities were adjusted to match measured salinity data. Adjustments were also made to achieve the appropriate residual flow in the model. An example of the circulation field predicted by the hydrodynamic model is presented in Figure 4.11.

Figure 4.10 Nested Finite Element Grid

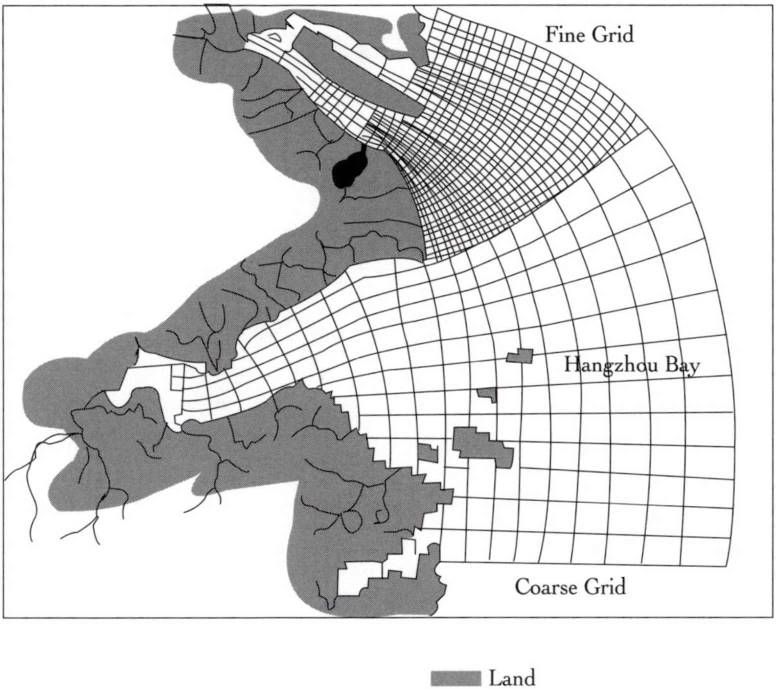

Delft used its water quality model called DELWAQ, which used the circulation pattern simulated by the hydrodynamic model as input. DELWAQ predicts cohesive suspended solid kinetics (sedimentation, resuspension, absorption, and mineralization), nutrient dynamics including nitrification and denitrification, dissolved oxygen processes, phytoplankton kinetics, heavy metals, and indicator bacteria. The water quality calibration process was similar to the hydrodynamic calibration and was carried out sequentially as follows: transport of suspended particles followed by total concentration levels, then water quality and ecological processes. Some of the model parameters adjusted in the calibration process included critical shear stresses; settling velocities; dissolved and detritus components of nitrate, ammonia, phosphate, and silicon; phytoplankton

Figure 4.11 Hangzhou Bay Simulated Flow Field

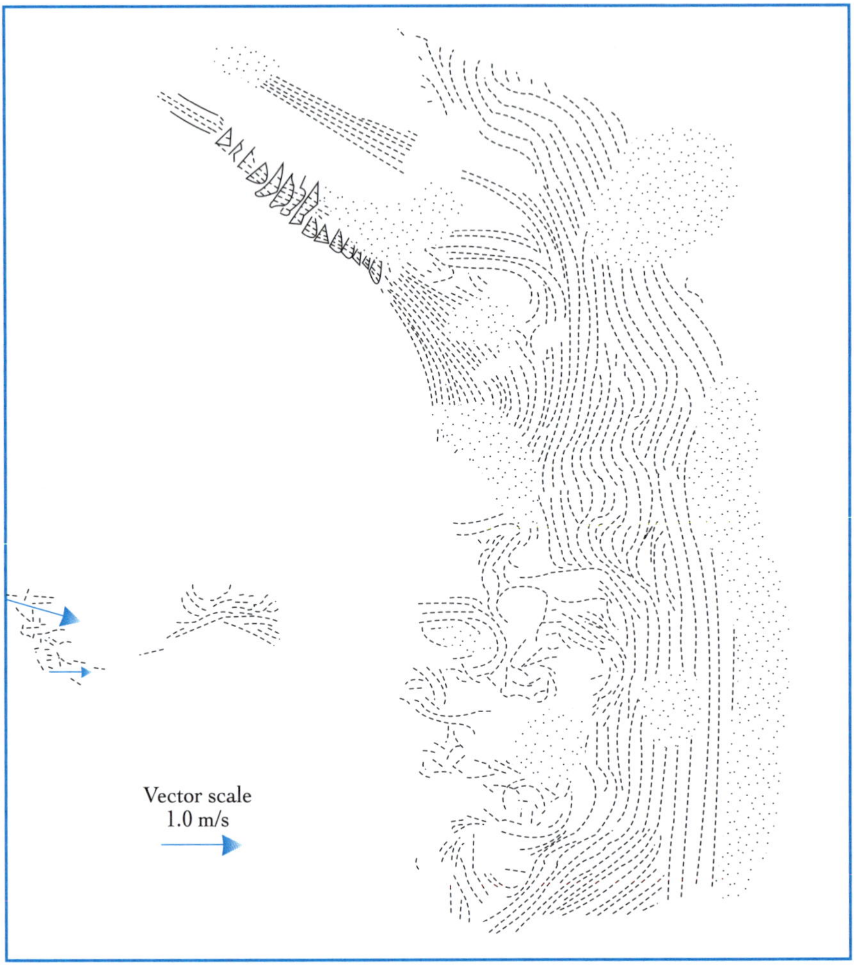

growth and mortality rates; nitrification and denitrification rates; mineralization rates; re-aeration flux; metal and carbon partitioning coefficients; and mortality rates for indicator bacteria. Figure 4.12 shows some of the results of the modeling predictions. The top figure shows the simulated freshwater fraction (Yangtze River water) in the bay, and the bottom figure shows the difference between simulated and measured salinity in the bay.

The modeling showed that the predicted increases in biological oxygen demand, indicator bacteria, and nutrients will not exceed

Figure 4.12 Hangzhou Bay Simulated Freshwater Fraction and Salinity Calibration

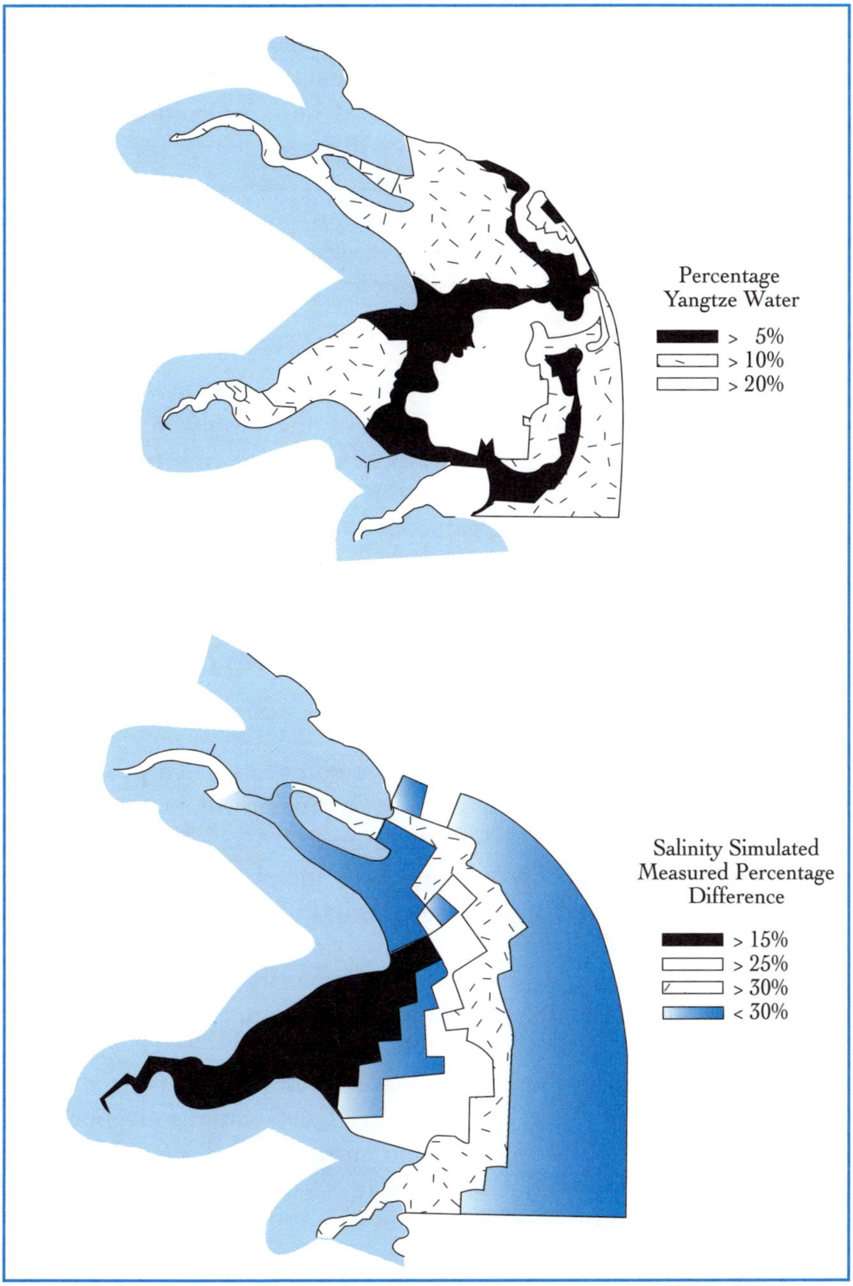

the assimilative capacity of the bay. However, because primary productivity is presently limited by high concentrations of suspended solids in the bay, any reduction in suspended solids concentrations will result in higher phytoplankton bio-masses, which could lead to greater occurrences of undesirable algal blooms because of the abundance of nitrogen available. Rural nutrient controls must be implemented in the river catchments if these blooms are to be controlled. The increased loadings of copper and zinc will result in increased sediment concentrations offshore on the order of 50 percent, which will have a accumulative effect on benthic organisms. Metal discharges in the river catchment areas need to be controlled.

Costs

Half a million tonnes of fish are annually harvested in the bay (10 percent of China's total production), and an evolving tourist industry on the islands is dependent on the water quality in the bay. The current annual commercial value of these resources is large. The water quality modeling cost relative to the annual commercial value of the resource was less than 1 percent.

SECOND SHANGHAI SEWERAGE PROJECT (SSPII), 1996

This project was the second stage in a comprehensive program for provision of wastewater management facilities for the city of Shanghai, China. The first stage had diverted 1.4 Mm^3/d of wastewater from the catchment of the Suzhou Creek in urban Shanghai for disposal to the Yangtze Estuary after preliminary treatment at a submarine outfall at Zhuyuan. The SSPII provides for the disposal of additional wastewater from the urban area, initially a discharge of 1.7 Mm^3/d, and ultimately 5 Mm^3/d, to the Yangtze Estuary at Bailonggong

The objectives of the project were to enhance wastewater and stormwater management by expanding wastewater collection, pretreatment, and disposal capacity and stormwater drainage facilities; to reduce urban pollution; to improve wastewater utility financial and operational management; and to strengthen sector institutions through training.

An environmental assessment was required to determine the possible effects on the local water quality in the Huangpu River and Yangtze Estuary.The environmental assessment confirmed that the assimilative capacity of the Yangtze River (mean annual flow 30,000 m^3/s) was such that in a far-field context, the discharge of the planned quantities of wastewater would not significantly adversely affect water quality in the immediate term. However, water quality modeling was necessary initially to assist in determining the optimal location of the outfall discharge at Bailonggong and the optimal diffuser configuration in order to minimize the near-field impact of the wastewater discharge.

The success of this project was ensured by the positive and cooperative working relationship between the Government of Shanghai and the technical staff from the East China Normal University and the Danish Hydraulic Institute.

Summary of the Project

Water quality modeling was required to determine the optimal outfall location, its design, and the wastewater treatment strategies, including the concept of phasing the project. In particular, water quality modeling was required to assess the nutrient loadings to the estuary and their implications in the frequency of occurrences of red tides in the estuary. The initial capacity of the outfall was 1.7 Mm^3/day, with an ultimate capacity of 5.0 Mm^3/day.

The water quality modeling was undertaken by the government of Shanghai, which engaged local Shanghai specialists from the East China Normal University and arranged for additional support from the Danish Hydraulic Institute. It was recognized at the outset that the circulation characteristics in the Yangtze Estuary are a complex function of the river flow, tidal stages, wind climates, and the Japan Current (Kuro Shio); consequently, the currents in the estuary vary in both time and three-dimensional space.

The DHI MIKE system of models was used for the predictions (for details, see Appendix) These models numerically solve the appropriate coupled partial differential equations on a rectangular grid. Three nested hydrodynamic models were used in this project, with grid sizes of 1,000m (M1), 250m (M2, and 40m (M3). The larger-scale models (M1 and M2) were two-dimensional, and the smaller scale three-dimensional. The areas modeled are shown in Figure 4.13. The results from the M1 model were used as boundary conditions for the M2 model, and the M2 results for the M3 model. The model inputs included bathymetry, bottom roughness, eddy coefficients (no site-specific data were available), tidal conditions, wind climates (shore-based measurements), and river flows. First, the M1 model was calibrated by comparison between the predicted water levels and the measured values for tidal water level variations and wind and net flow conditions. A similar calibration process was carried out for the M2 and M3 models. In addition, the calibration of M2 model included the comparison between the simulated and measured current velocities.

The water quality process for quantifying the required model input parameters like dispersion, biological oxygen demand, indicator bacteria decay coefficients, and suspended solids partitioning

Figure 4.13 The Model Domain

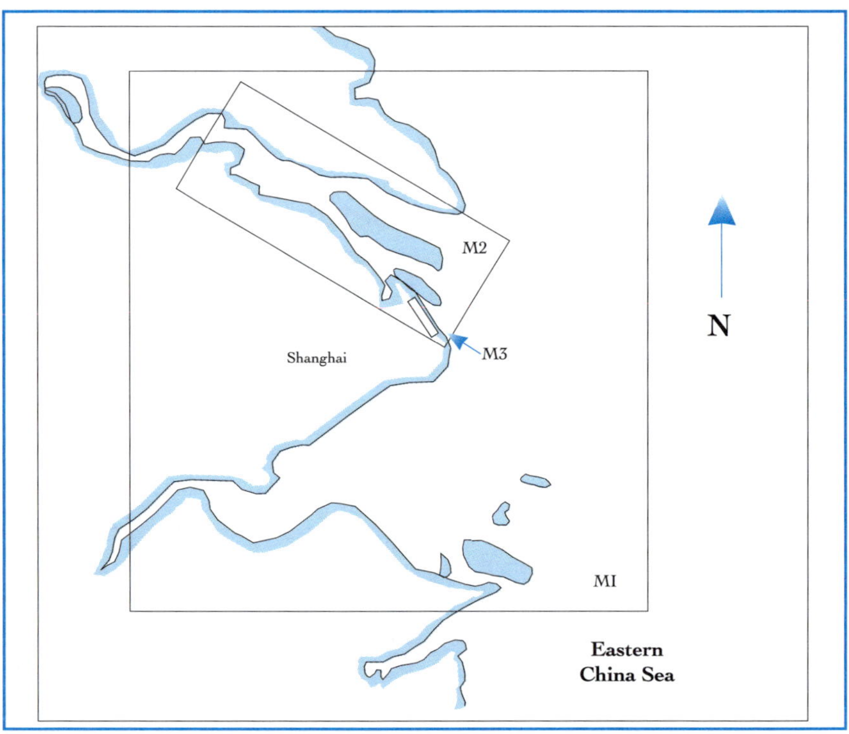

coefficients for the heavy metals is discussed. The M2 water quality model was calibrated for indicator bacteria, biological oxygen demand, dissolved oxygen, ammonia, nitrate, and phosphate over a 7.5-day period for both a dry season (March) and a wet season (July). In the calibration process, the decay rate and temperature coefficients, nitrification and denitrification constants, nutrient contents of organic matter, and suspended solids partitioning coefficients for metals as well as the sedimentation and critical resuspension velocities were adjusted to produce predictions that were similar to the measured values.

The models were used to assess different abatement options, with emphasis on comparing modeling predictions for different options rather than comparing the model predictions with water quality objectives or standards (see Figure 4.14). The model predictions showed that the assimilative capacity of the Yangtze River Estuary is mainly dependent on the river discharge, tidal variation, background concentration, and biochemical processes in the estuary. The goal of the outfall design and location study was to optimize exploitation of this assimilative capacity, which is a direct function of the water quality standards for the estuary. The simulated results confirmed that 1.7 Mm^3/day of wastewater could be discharged at Bailonggong without significant adverse effect on the water quality in either the near or far field. The optimum diffuser configuration, recommended in light of the modeling results, ensured complete mixing within a limited area in the near field and comprised a 200m-long multiport diffuser consisting of six risers. For aesthetic purposes, it was recommended that oils be reduced in the effluent.

The water quality modeling was also used in the development of a longer-term strategy for wastewater treatment for Shanghai, in particular the need to control nutrient loadings to the Yangtze Estuary. This was important because nutrient loads affect the frequency of red tides in the nearby Hangzhou Bay.

Costs

The capital costs for the overall project were estimated to be US$633 million. The governments of Shanghai, Denmark, Norway, and France provided the US$450,000 modeling costs. The modeling had a substantial economic and financial benefit because it provided

Figure 4.14 Simulated Near-field Surface Concentration Distribution of Copper

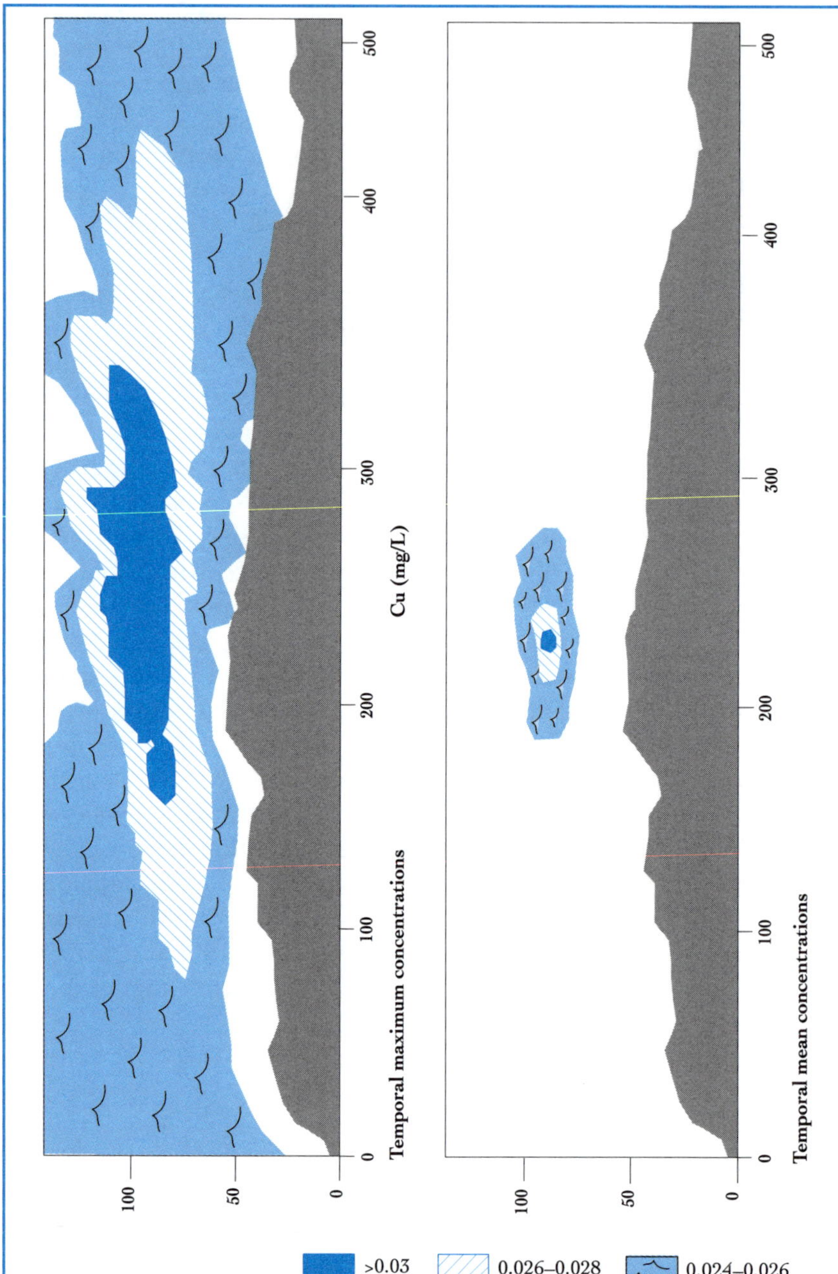

direction for development of a phased approach to wastewater management.

SHANGHAI ENVIRONMENT PROJECT, 1994

The population of the Shanghai municipality in 1991 was about 14 million, with some 8 million living in the urban center. Most of this population was served by water drawn from the Huangpu River. The river has played a major part in the development of the city, both as a route for some 7,000 shipping movements per day between the coast and the Grand Canal and as a source of water for the many factories which have been built beside the river. The quality of the water in the river has been degraded by both municipal and industrial discharges. The seriousness of the situation was appreciated by the Shanghai government in the early 1980s; consequently, an intake was built, upstream of the existing city intakes, at Linjiang, with raw water fed to the existing treatment works by an enclosed conveyor. However, even this improvement did not allay concern regarding the variety of industrial discharges upstream of Linjiang and those washed upstream to Linjiang on rising tides. It was therefore decided to extend the conveyor further upstream. The goal of the project was to determine where the new intake should be located to provide a safe drinking water supply.

Summary of the Project

The Shanghai area is very flat, and areas of the municipality are below high tide water levels and protected by the Bund; therefore, tidal effects are important. It was decided that a one-dimensional model designed to characterize tidal effects was appropriate to examine the suitability of a new site at Da Qiao.

The water quality parameters modeled were BOD, ammonia, nitrate-dissolved oxygen, and phenols. Phenols were included because there are high concentrations of phenols in the river and because there was an accidental spill of large quantities of phenols in 1989.

Rather than simulate a long period of historical records, it was decided to select a particular set of hydraulic conditions that would represent a worst-case scenario, and each model run examined only two tidal cycles. In this way, it was possible to carry out a large number of trial runs in a short period of time.

The natural river flow is largely regulated by controlled discharges from Lake Tai. It was therefore possible in principle to reduce the impact of upstream tidal effects by controlling the flow of the river at a high level. This approach was tested by using the mode, as were other scenarios, including different sewage and treatment options.

The findings of the modeling process were as follows:

- A paper mill discharging to the river upstream of Da Qiao should be closed.
- Other discharges upstream of Da Qiao were at acceptable levels.
- Normal discharges from existing factories downstream of Da Qiao were having a negligible negative effect upon the quality of water at Da Qiao.
- Simulation of the phenol spill referred to above revealed that at normal river flows the impact of the incident would be detected at Da Qiao.
- The effect of increasing the river flow to 300 m^3/s would be to eliminate the risk of a pollution incident in the industrial areas of Minhang and Wujing that would adversely affect the water quality.

The findings of the modeling process resulted in the following:

- The paper mill at Zhongjiang was relocated.
- An environmental protection zone was established for the length of the river with a width of 10 km. No development in this zone is permitted without approval by the Shanghai Environmental Polution Bureau.
- A new intake was constructed at Da Qiao, with a new conveyor of 5.4 Mm^3/d capacity delivering water to the existing conveyor at Linjiang.
- Sewage from the Minhang/Wujing areas was given high priority, and consideration for its potential adverse effects was given in all works constructed under the SSSPII. Thus, failures in treatment or acciden-

tal spills in the industrial area will not affect the quality of the Huangpu River but will be discharged to sea.

Since the operation of the new intake, the quality of drinking water in Shanghai has noticeably improved, and with the sewerage works referred to above completed, it has not been necessary to augment the flows of the Huangpu.

Costs

The options evaluated to solve the drinking water quality problems were as follows:

- Relocation of the intake — Cost – US$253 million
- Advanced wastewater treatment — Cost – US$446 million
- Withdrawal of water from the Yangtze — Cost – US$1058 million
- Improved local pollution control — Cost – US$760 million

The cost of the modeling was about US$200,000; without it, the cheapest option, relocation of the intake, could not have been selected. This sum represents less than 0.1 percent of the project cost and less than 0.1 percent of the potential savings.

Conclusions

The highly satisfactory work was largely the result of excellent cooperation between the Shanghai government, in particular the staff of the SEP Project Office, and Mott MacDonald, the international consultants who were funded by the UK government. Of particular help were specialists from the He Hai University, who brought amazingly detailed knowledge of the hydraulics of the myriad waterways linking Lake Tai and the Huangpu River.

Lessons to be learned from this most successful case are as follows:

- Define clearly the objectives of the modeling.
- Keep the models as simple as possible.
- Use specialists to make the assumptions and judgments.

MANILA SECOND SEWAGE PROJECT, 1996

To improve the quality of waterways in Metropolitan Manila and Manila Bay, the Philippines, and to reduce the health hazards of exposure to human wastewaters, water supply and sewerage services needed to be upgraded. Eighty-two percent of the residential population of Manila use septic tanks for their wastewaters. The septage from these tanks will be treated in land-based facilities which are proposed to be constructed during the period 2003-2010. As an interim measure to reduce the human health risks of exposure to wastewaters in the metropolitan area, it was proposed that the septage be dumped in the open ocean. The locations of the ocean dumping, the methods of dumping, and the schedule of the dumpings had to be developed in such a way as to comply with the 1972 (with subsequent amendments) "International Convention on Marine Dumping of Wastes and Other Matter" and to minimize the negative effects of the dumping on the receiving waters. Initially, 1,000 m^3 of septage will be dumped in the ocean per day. This volume was expected to increase to 1,500 m^3 by year 2001, then decrease as the land-based treatment facilities commence operation.

Terms of reference for a water quality modeling study were prepared to identify the best locations for the dumping and the methods of dumping and to examine the potential effects of the ocean dumping on coastal areas and recreational beaches. One of the loan requirements was that environmental impacts be predicted and, if these impacts were deemed unacceptable, the project design would have to be changed or other mitigation procedures undertaken to abate the impacts.

Water Quality Modeling Requirements

A hydrodynamic model was required to simulate the three-dimensional circulation in Manila Bay caused by surface winds, tides, and density gradients for different seasons. The area for the simulation had to include the proposed dump sites and the shoreline North of Manila Bay (about 90 km north). The hydrodynamic model was to be used to predict the transport, dispersion, and fate of dissolved and particulate pollutants from the dump sites for various loadings.

Concentrations and zones of settlement were to be predicted and compared to international environmental quality standards. Cumulative effects of dumping at the rates proposed were to be investigated. The modeling contract was awarded to the Danish Hydraulic Institute.

Summary of the Project

The model used for this project, MIKE 3 PA, consisted of four elements satisfying the conservation of mass, momentum, salinity, and temperature and the equation of state relating local density to salinity, temperature, and pressure. A brief description of the model is presented in the Appendix. The hydrodynamic model was a fixed grid model, and equation solutions were obtained numerically. The main input data for the model consisted of boundary conditions (two of four boundaries were open, with no data available), topography, bathymetry, wind climates (shore-based meteorological stations), bottom frictions (no site-specific data were available), loadings, material specifications, dispersion and decay coefficients (no site-specific data were available), velocity depth profiles specifications (no site-specific data were available), and simulation period. The main two-dimensional model outputs were instantaneous and averaged circulation vectors, particle concentrations, and erosion/deposition/net sedimentation. Figure 4.15 shows the model grid and surface circulation vectors. Calibration of the model parameters used the available water-level data from two sites; meteorology data available from shore weather stations; and Coast Guard data on currents and Manila Bay temperature, salinity, and water quality. Because the site-specific data were limited, the model was used to compare scenarios for the same conditions. Figure 4.16 shows these comparisons. Verification of the model predictions was not possible because of the lack of site-specific data: consequently, the quantitative model predictions are not rigorous.

The model predicted that the depth of solids deposition on the ocean floor would be negligible (0.04 mm for the three-year period, for a total of 58,950 tonnes of solids) and therefore would not affect the fishing grounds. The nutrients provided from the dumping could affect the primary productivity in the vicinity of the dump

Figure 4.15 Simulated Current Velocity Vectors

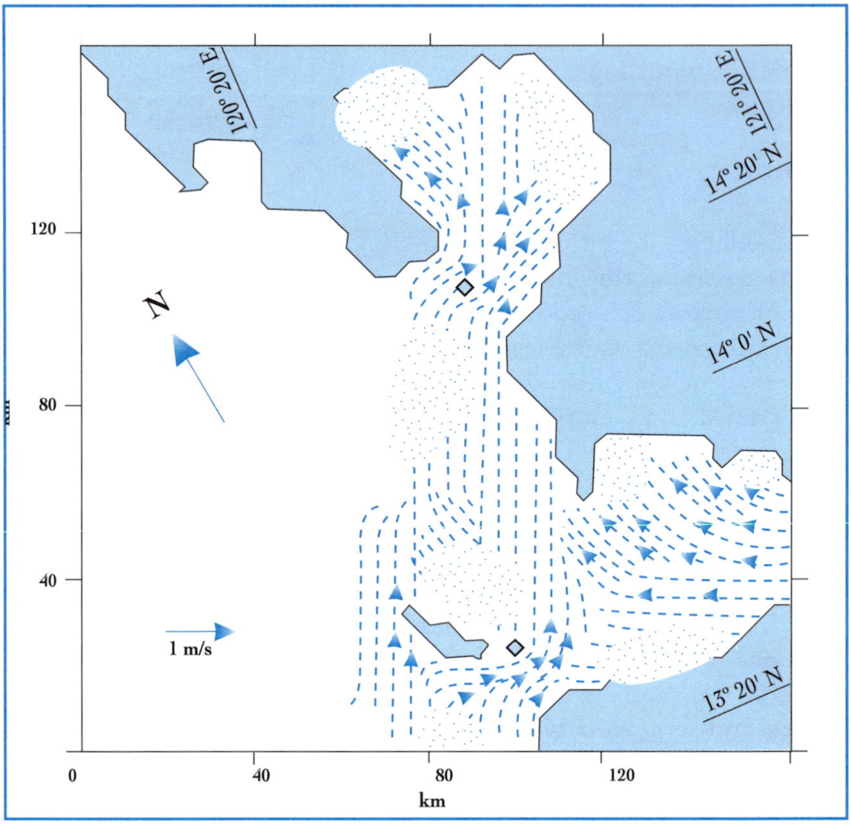

site. The impact of the dumping will be monitored by measuring turbidity with a Secchi disk. In some instances, bacterial densities at the recreational beaches and coastal area to the northeast of the ocean dumping site were slightly elevated, and a monitoring program was recommended for those areas. Spills from the barges could also degrade the bacterial quality in the coastal areas and recreational beaches. The spill potential was reduced by restricting barges to the shipping lanes and periodic barge inspections.

Figure 4.16 Simulated Benthic Loadings

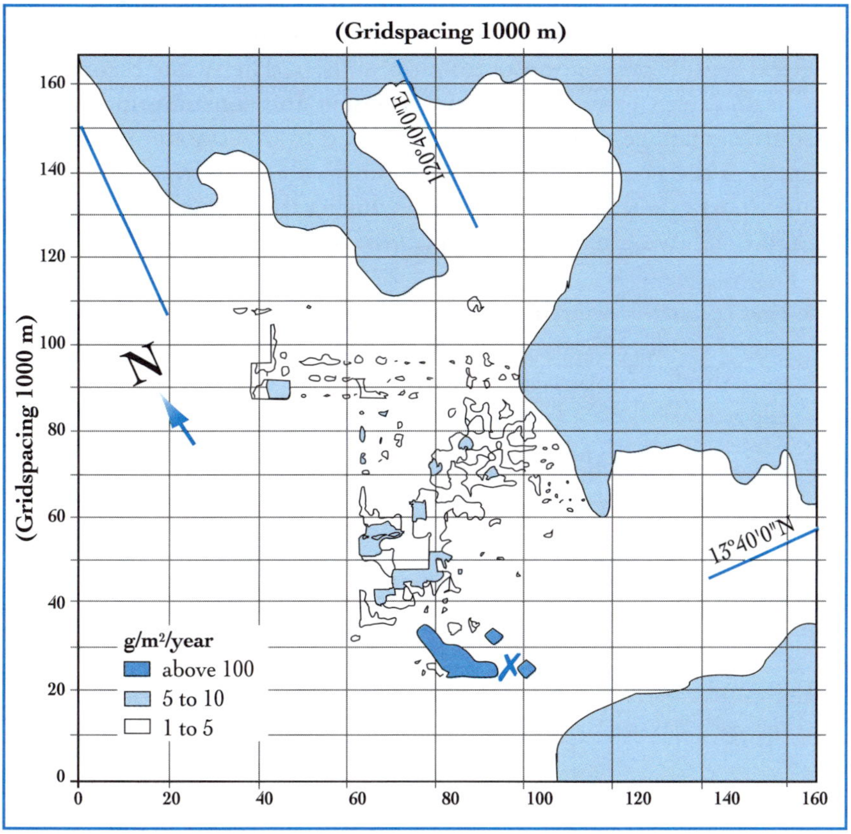

Costs

The cost of the water quality modeling was less than US$100,000 in 1996. If the ocean dumping of septage had been an acceptable option, the land disposal option would have to have been used The landfill option would have cost US$29.4 million; the ocean dumping, by contrast, cost US$19.4 million. Financing for the modeling was provided by the Danish Consultant Trust Fund.

TARIM BASIN II PLANNING PROJECT, 1997, CHINA

The Tarim Basin in northwest China is more than 600,000 square kilometers, with a population of about 6 million. The water resources in the basin are extensively used for irrigation and consumption A water resources management study was initiated to investigate methods for managing the water resources more effectively, in particular to relieve water shortages that tended to occur during the springtime. In particular, the following management proposals were to be assessed:

- land reclamation and land improvement, including improvement of irrigation system efficiency through land leveling and other measures, and the introduction of surface drainage for water table control and salt export;
- improved canal lining, which would result in a reduction in seepage losses from canals;
- improved on-farm management;
- use of surface reservoirs for resource regulation (although use of this option was discouraged); and
- groundwater abstraction, to alleviate water supply shortages during the spring period and for drainage purposes.

Water Quality Modeling Requirements

Water and salt balance models were required to assess the results of the various development options. The objectives for the model development were as follows;

- to relate the impact of the development options on the surface water contributions to the Tarim River to the reduction of non-beneficial losses within the sub-basins and to the salt balance;
- to determine how surface water drainage would affect the salt balance and how it might reduce non-beneficial losses of salt in the sub-basins;
- to prepare a framework for more detailed modeling of the integrated surface water and groundwater systems within the sub-basins in the next phase of the study; and

- to make recommendations for future data collection, monitoring, and river modeling.

Summary of the Project

The models were developed and tested under contract to the Mott MacDonald consulting firm and specialists from the Xinjiang Agricultural University, Xian University, and Tsinghua University. Figure 4.17 is a map of the region for which the models were developed; however, each sub-catchment also had a separate model. The overall model development components are shown in Figure 4.18. In general, the water balance model consisted of river, reservoir, irrigation, groundwater, and lake modules, with the model formulation in each module being different. For example, the river module consisted of inflows, outflows, diversions, exports, gains and losses to groundwater, evaporation, return flows, and storage changes. The lake module consisted of inflow from drainage surpluses, inflow from the river module, evaporation, inflow and outflow to groundwater, change in lake storage, and other specified outflows. The salt balance model consisted of reservoir, river, irrigation/groundwater, irrigated and non-irrigated fields, and lake modules. Like the water balance models, the salt balance models are formulated differently in each module. The models were calibrated "to reach an acceptable agreement between the observed and simulated water balance components." Figure 4.19 shows the results of calibration. Once the models were calibrated, model prediction runs were carried out as follows:

- *Sensitivity analysis.* The model parameters were systematically changed and the effects of these changes on the individual water balance components analyzed. Specifically, the sensitivity analysis identified those water components that were important and established uncertainty ranges for the controlling parameter to test the impact of a change to one or several parameters on the water balance.

- *What-if scenarios.* Model runs are used to obtain a better understanding of the benefits of certain assumed changes to model parameters on the water balance. What-if scenarios were used in the runs. For example, "If I were to increase the depth to the water table, what would be the effect on capillary losses from the water table and what would be the

Figure 4.17 Tarim River Basin: Stage II Project Location

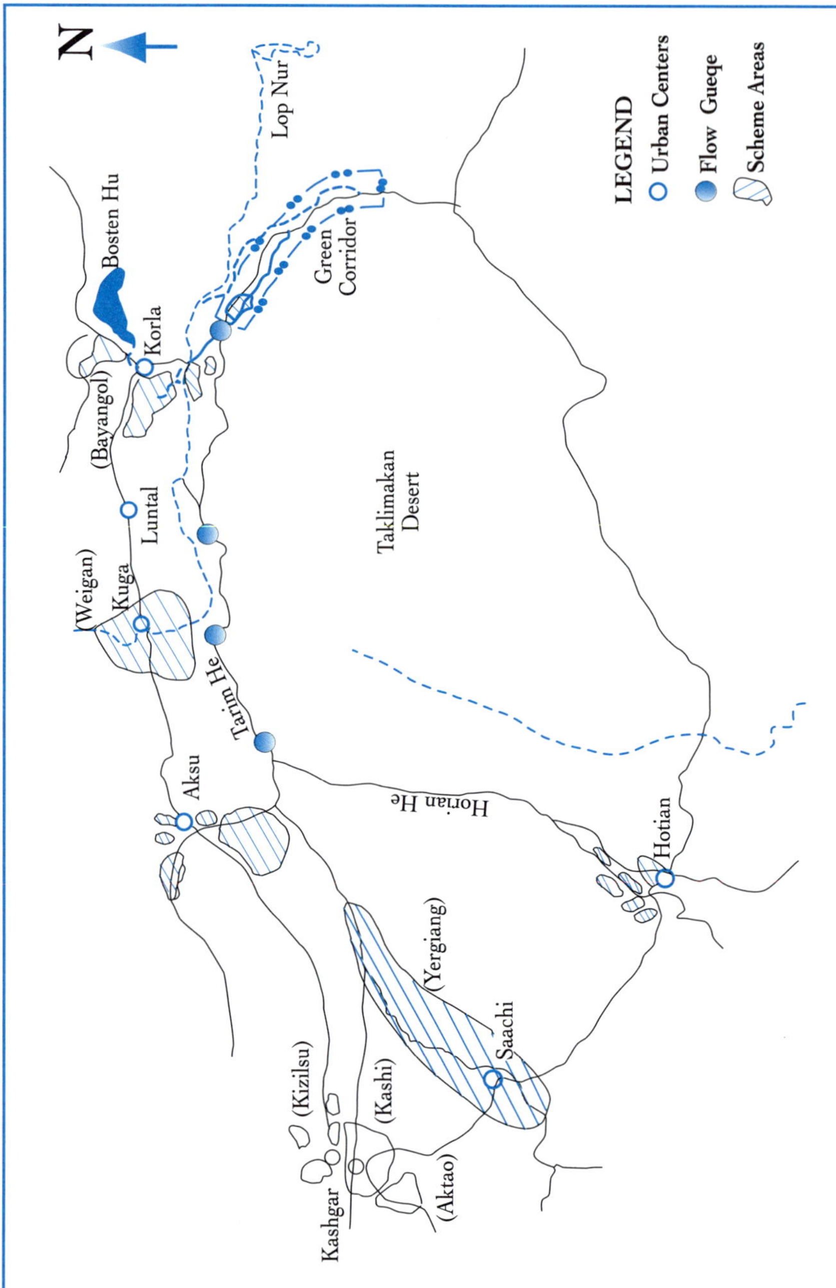

Figure 4.18 Tarim II Preparatory Study: Study Activities

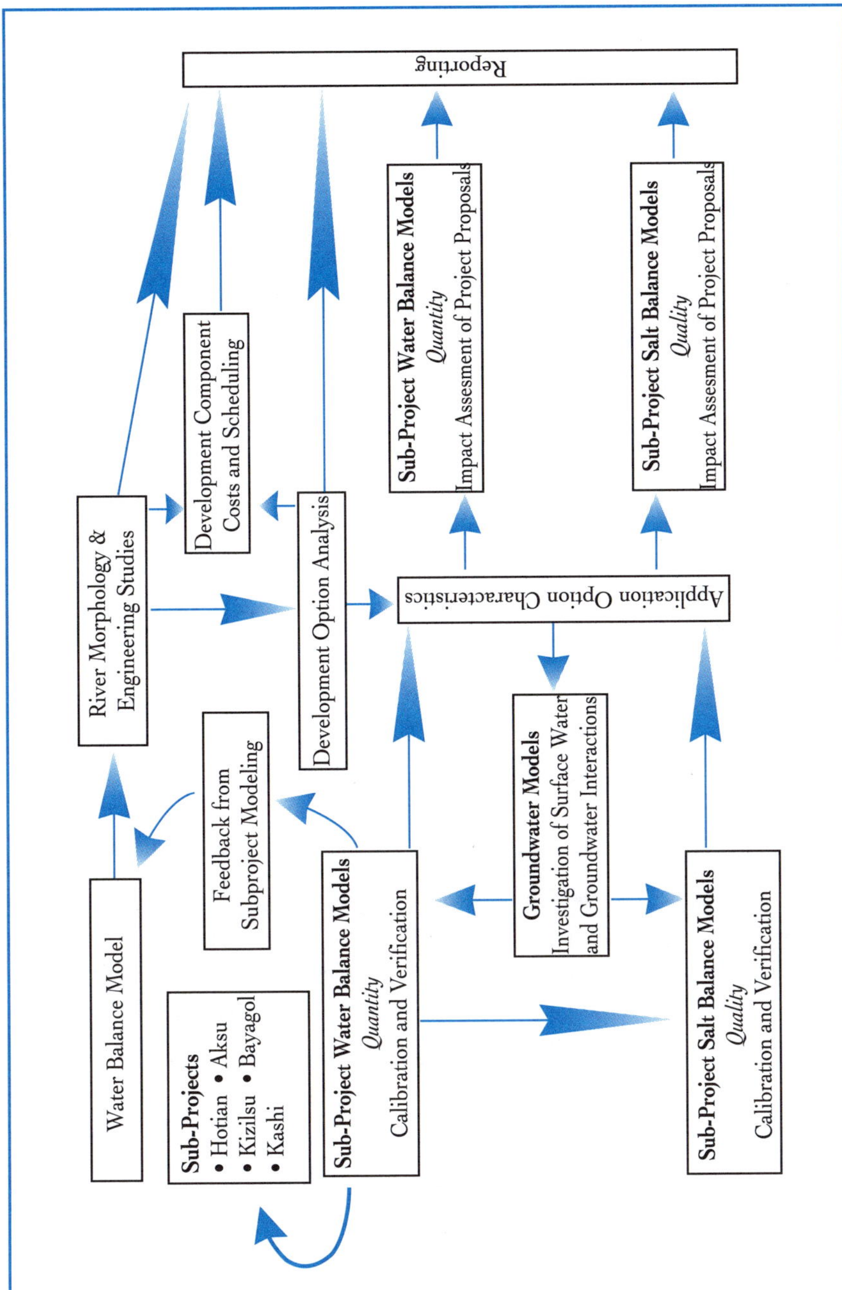

Figure 4.19 Simulation of Bostan Lake

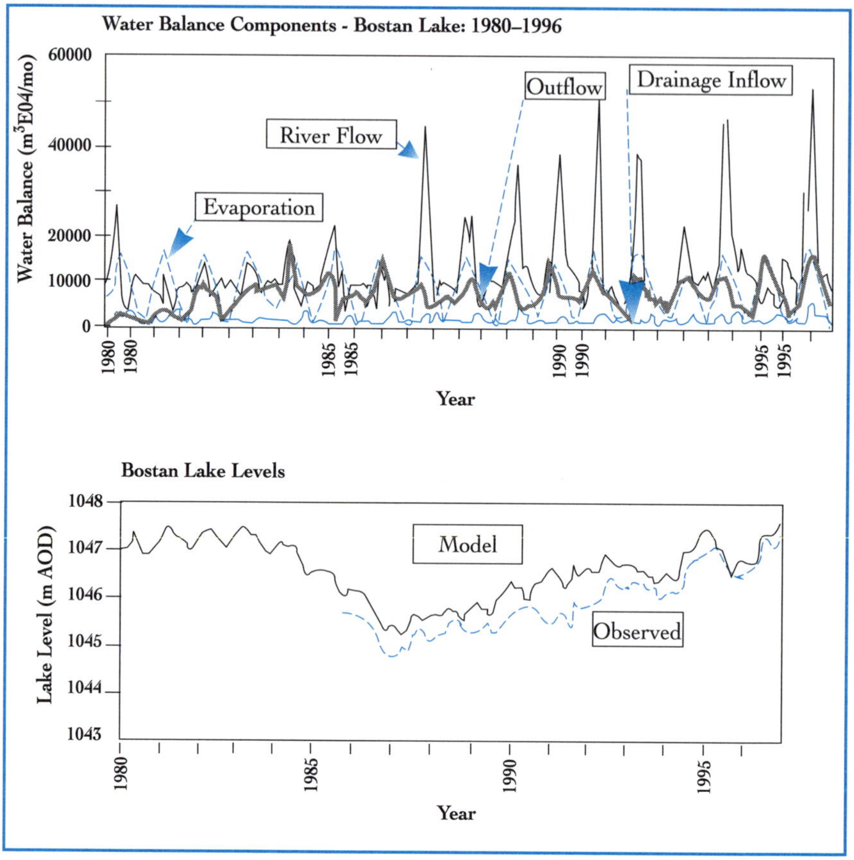

broader impact on the water balance?" "What will be the benefit of improved canal lining and irrigation efficiency on the availability of the resource?" "Would there be any benefit if the water table in an area where a wellfield is planned were relatively deep?"

- *Model predictions for the project as required.* Because there is a need for many model runs, good recordkeeping and an audit trail are essential.

The models were used to assist in deciding which canals should be lined, where wellfields should be placed, where drains should be installed, and where low-yield land improvement and land reclamation activities should be undertaken. The design of the Tarim Basin

II Project (which is presently under implementation) was determined in part on the basis of the model results.

In addition, modeling requirements for the next phase were identified as well as the monitoring and other data collection requirements to support the prediction models.

Costs

The modeling activities cost approximately US$80,000, which was financed with a Japanese PHRD preparation grant for the Tarim Basin II Project.

Appendix

CE-QUAL-W2: A NUMERICAL TWO-DIMENSIONAL LATERALLY AVERAGED MODEL OF HYDRODYNAMICS AND WATER QUALITY

Source: Environmental Laboratory
U.S. Army Corps of Engineers
3909 Halls Ferry Road
Vicksburg, MS 39180-6199
Tel: 601-634-3283
Program and manual are available on the internet. Contact the above for access information.

Model Hardware Requirements: A recent personal computer with 32K RAM and 2Meg hard drive with math co-processor with input disk file and line printer or disk for output.

Professional Expertise Requirements: A university science degree with experience in hydrodynamic modeling and water quality modeling.

Receiving Water: Primarily narrow reservoirs.

Spatial Characteristics: Two-dimensional laterally averaged.

Temporal Characteristics: Dynamic.

Water Quality Parameters Simulated: Dissolved oxygen, biochemical oxygen demand, nutrients, algal bio-mass, temperature, indicator bacteria, conservatives and non-conservatives as well as the interaction of flow and temperature and nutrients and algal productivity.

Model Applications: Reservoir management for release times and volume rates to maintain water quality in the reservoir.

Model Attributes

This model predicts the water quality processes in a temporally density stratified reservoir. The model predicts the surface elevations, velocities, temperatures, and substance concentrations as well as the downstream release concentrations. The input data for the model includes the reservoir topography, bathymetry, input and output flows, atmospheric heat flux parameters, initial conditions, and the numerical grid parameters and time steps. The water quality modeling processes are similar to those used in QUAL2 and WASP but, unlike these models, the reservoir density stratification due to water temperature is dynamically predicted. Density stratification affects the water quality processes in the reservoir. These processes are predicted in this model.

Physical Aspects of the Model

The physical characteristics of the model are similar to those of WASP, except that CE-QUAL-W2 is laterally averaged and will predict the thermal density depth gradients dynamically.

Model Error and Sensitivity Methods

The model error and precision must be determined by the model user.

General Comments

This model is based on an earlier reservoir model that has been enhanced by using the water quality processes of QUAL2. This has made this model a very powerful prediction model for reservoirs and reservoir flow management. The model has been extensively used in North America.

CORMIX

Source: U.S. Environmental Protection Agency
Office of Research and Development
Environmental Research Laboratory
960 College Station Road
Athens, Georgia 30605-2700
Send two blank 3.5" HD diskettes. No other costs.
Manual available from NTIS.
Tel: 703-605-6000, ask for EPA/600/3-90/012, February 1990.

Model Hardware Requirements: Modern PC with mathematics coprocessor, input disk file, and line printer or disk for output.

Professional Expertise Requirements: A university degree with courses in hydrodynamics.

Receiving Water: Rivers, lakes, estuaries, and marine coastal areas.

Spatial Characteristics: One- or two-dimensional.

Temporal Characteristics: Steady state – does not have time as a variable; however, model can be run repeatedly to simulate time variability.

Water Quality Parameters Simulated: The initial mixing (mixing zone) of all water quality parameters.

Model Applications: All submerged outfalls, either single or multiports.

Model Attributes

This model is an expert system that predicts the path and dilution characteristics of a discharged effluent plume for all possible conditions. These conditions include positive and negatively buoyant plumes, trapped plumes, bottom-attached plumes, unstable plumes, as well as conventional plumes. Figure CRX 1 shows the scope of the different conditions that can be simulated in CORMIX. The equations used to predict the plume trajectory path, cross-sectional area, and dilution for each case are presented and discussed in the user's manual. This model is suitable for all submerged outfalls regardless of the receiving water. For a particular application, the user supplies information about the discharge and ambient environment. The model returns information detailing the hydrodynamic mechanisms controlling the flow, dilution, geometric information concerning the shape of the pollutant plume, or flow in the ambient water body, as well as design recommendations, allowing the user to improve the dilution characteristics of the flow. Figure CRX 2 shows an example of the model simulation and field observations.

Typical model runs take about five minutes, and a user will require two to three days working with the model to understand the various functional relationships in the expert system.

Model Error and Sensitivity Methods

There are no discussions of model error and sensitivity methods in the manuals. The user will have to develop methods for quantifying the prediction precision for a particular application. This would include varying the user input data, either singly or in groups for the range expected in the particular application. Determining the model precision could occur as part of the learning process in using the model.

General Comments

Because the user must provide information on the ambient receiving water conditions to the model, it is important that site-specific data be available. In particular, the range of values for the input data should be known. Depth, ambient receiving water currents, and ambient depth stratification are the most important input data. The model is very useful in the design of multiport diffusers and the determination of the best location and depth for the outfall.

Figure CRX 1 Flow Classification System

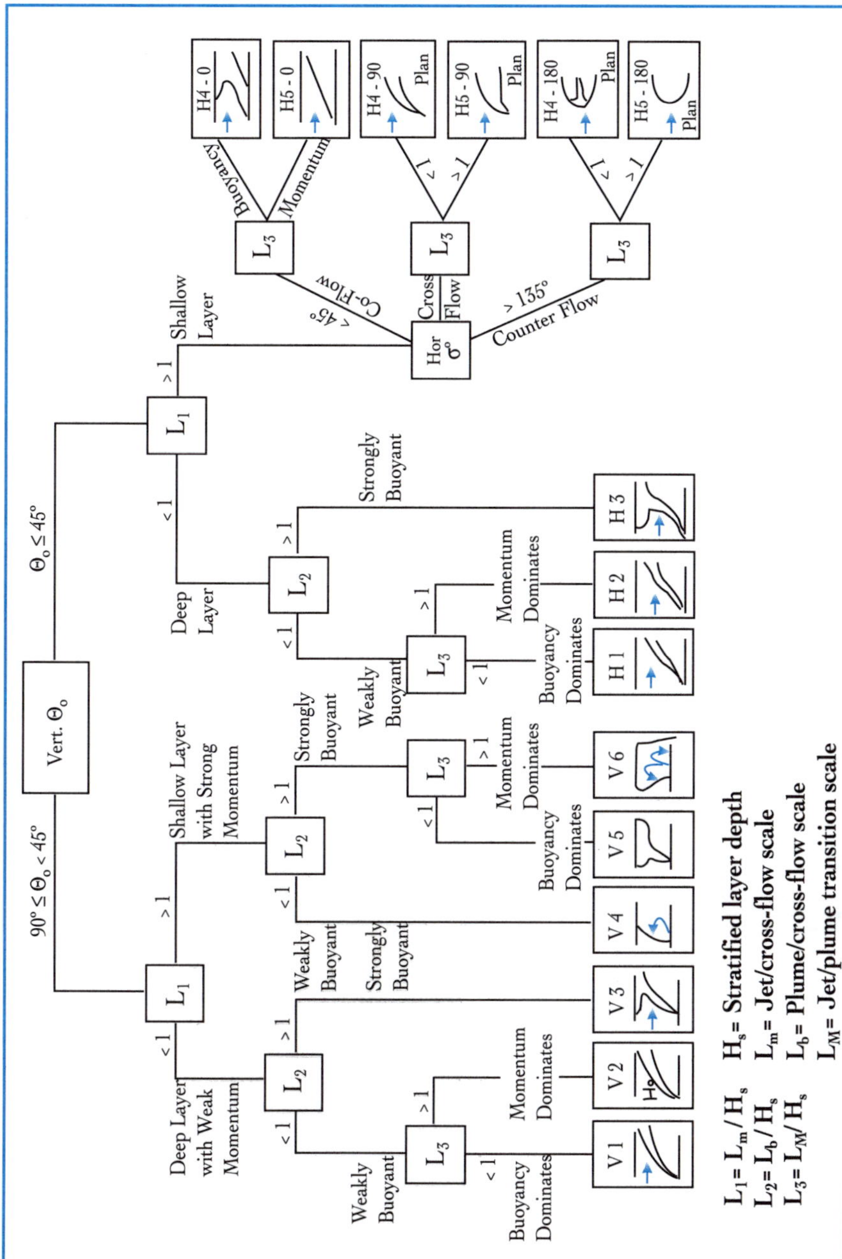

Figure CRX 2 Predictions versus Measurements

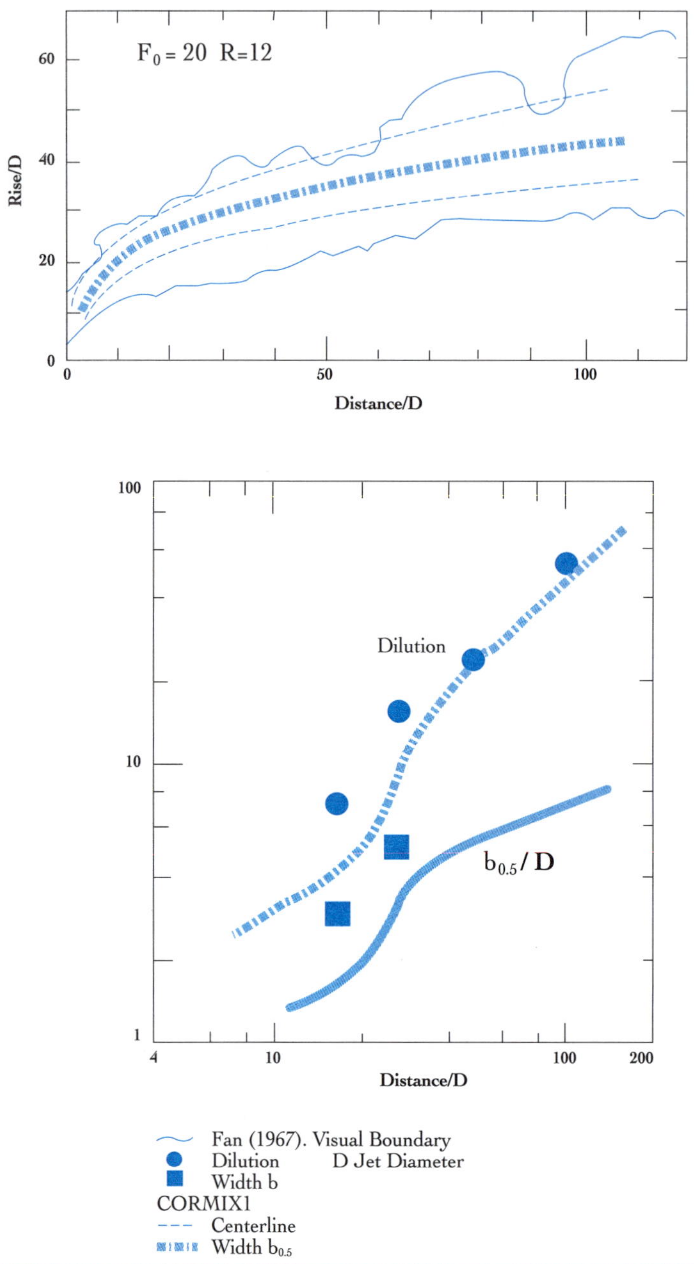

DIVAST
Binnie & Partners

Source: Binnie & Partners
Grosvenor House
69 London Road
Redhill, Surrey
United Kingdom
Tel: 44-1737-774155
Fax: 44-1737-772769
Details on the model are available at the address above.

Model Hardware Requirements: Unknown because it is a proprietary model, but a recent personal computer with 32K RAM and 2Meg hard drive with math coprocessor would be appropriate.

Professional Expertise Requirements: User should have a science degree and be knowledgeable in modeling techniques and have an appreciation of modeling physical processes.

Spatial Characteristics: Two dimensions.

Temporal Characteristics: Dynamic.

Water Quality Parameters Simulated: DIVAST (Depth Integrated Velocity And Solute Transport)—Soluble substances, including indicator bacteria.

Model Applications: Wide rivers, lakes, estuaries, and coastal regions.

Model Attributes

This model consists of the dynamic partial differential hydrodynamic equations in two dimensions, continuity equation, and the mass balance equation. These equations are coupled and solved numerically on a horizontal grid with variable cell sizes using an alternating direction implicit scheme (Figure 4.6). The outputs from the model include water surface elevation, current speed, and direction and concentration. The hydrodynamic model will accept continuously recorded wind, water depth, and current data. The model grid requires the topography and bathymetry. The continuously recorded wind, water depth, and current data near the model boundaries were used as input boundary conditions for the model calibration. In the calibration process, the bottom roughness coefficients, surface wind drag coefficients, eddy coefficients, and model time steps are adjusted so that predicted water depths and currents match continuously recorded depths and currents at interior locations on the grid (Figure 4.7). The solute calibration consisted of adjusting the dispersion coefficients to match measured bacterial densities. Site-specific data from tracer studies on the rising and falling tides and laboratory measurements of the bacterial mortality rates were also used in the model calibration (Figure 4.8).

Model Error and Sensitivity

Numerical dispersion can be computed from the grid size, velocity, and time steps used. The model predictions of depth, currents, and bacterial densities at various locations on the prediction grid are compared to continuously recorded data in the calibration process; consequently, the precision of the model predictions should be reasonable. The sensitivity of the model can be determined using another data set for verification. If continuously recorded data are not available, a sensitivity analysis can be carried out on the required data input parameters.

General Comments

The model was used to assess different locations and levels of treatment for two large outfalls for Bombay. The modeled area had the shore as one boundary, and the other three boundaries were open coastal waters. Numerical prediction models for a coastal area with three open boundaries are very difficult; therefore, Binnie & Partners used extensive site-specific continuously recorded data for the model development.

HYDROLOGICAL SIMULATION PROGRAM-FORTRAN (HSPF) User's Manual for Release 8.0

Source: U.S. Department of Commerce
National technical Information Service
5285 Port Royal Road
Springfield, VA 221161
Tel: 703-605-6050
Reference: EPA/600/3-84-066, June 1984
or
Environmental Research Laboratory
U.S. Environmental Protection Agency
960 College St. Road
Athens, Georgia 30605.
(706) 355-8000
Paper copy of the user manual is available from either address above. The program is free of charge from the Athens address above, but the request must be accompanied by six HD 3.5" disks.

Model Hardware Requirements: RAM 512 Kbytes, diskette, hard drive with more than 3Meg, numeric co-processor, DOS, and ANSI FORTRAN.

Professional Expertise Requirements: User should have a science degree, be knowledgeable in modeling techniques, and have an appreciation of the modeling physical processes. HSPF is a large, relatively sophisticated hydrologic hydraulic and water quality simulation program that does not include a sewer network.

Spatial Characteristics: Two dimensions.

Temporal Characteristics: Dynamic.

Water Quality Parameters Simulated: Dissolved substances, suspended solids, dissolved oxygen, nutrients, and indicator bacteria. The water quality parameters are linked to the runoffs generated in the hydrological modules.

Model Applications: Predicts runoff from rainfall and topographical information for the catchment.

Model Attributes

This is a complex model that predicts runoff in real time from rainfall data. The SWMM extended HSPF's capabilities to urban areas. SWMM is similar to HSPF in structure. HSPF has many modules that the model user can select for a particular application. The model has been extensively used to predict runoff and runoff quality. The manual contains extensive information on the receiving water quality processes; however, these processes are common to most models. Like the SWMM model, HSPF can be used as a black box runoff prediction model with an abbreviated input data set if calibration and verification data are available.

Model Error and Sensitivity

The accuracy of the runoff prediction from HSPF should be similar to SWMM's. The error can be determined by using verification data and/or using sensitivity analysis for the user provided coefficients and rate constants.

General Comments

HSPF is one of the early runoff prediction models developed. The quality components of the model are all linked to the washoff or runoff components of the model. The model has a good suspended sediment kinetics module (Figure HSP 1 from the user's manual). The nitrogen and phosphorus processes considered in the model are shown in Figures HSP 2 and HSP 3, respectively. There is also a reservoir quality module (Figure HSP 4). While it is possible to select the appropriate model module, the hydrological module outputs must be available. This model is useful for river basins, which are largely rural.

Figure HSP 1 Flow Diagram

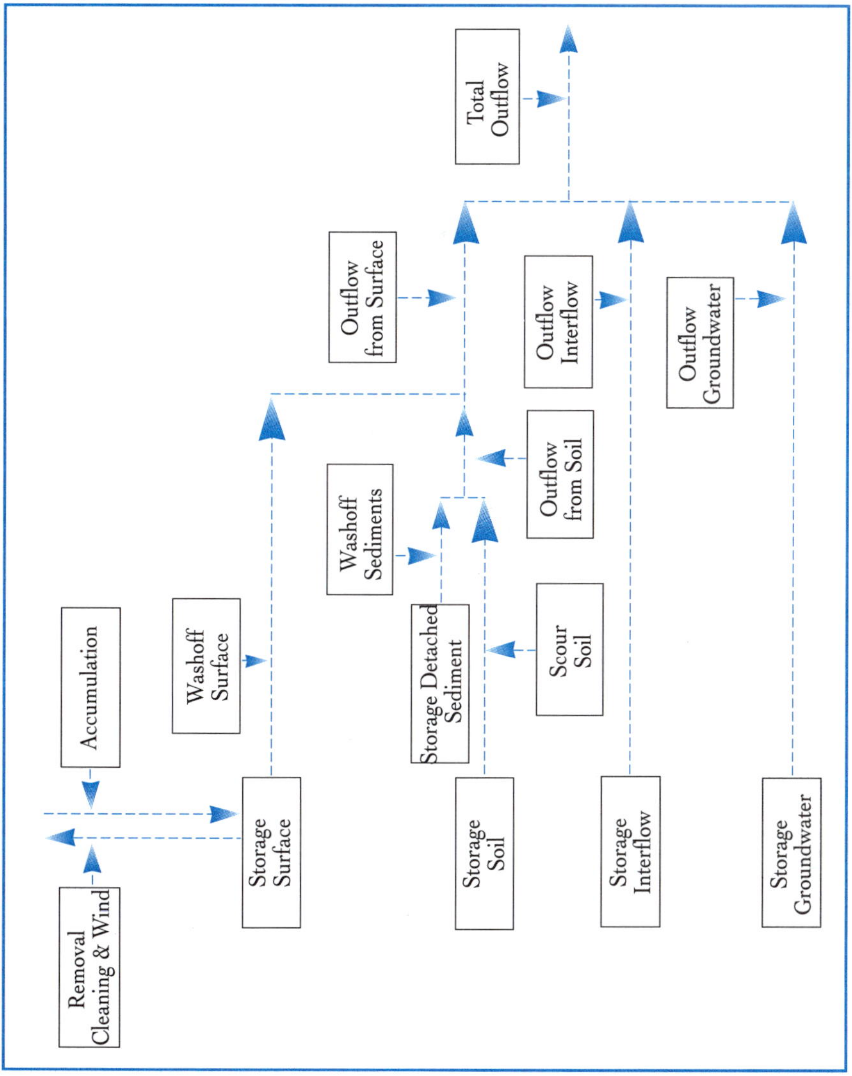

Figure HSP 2 Flow Diagram for Nitrogen Reactions

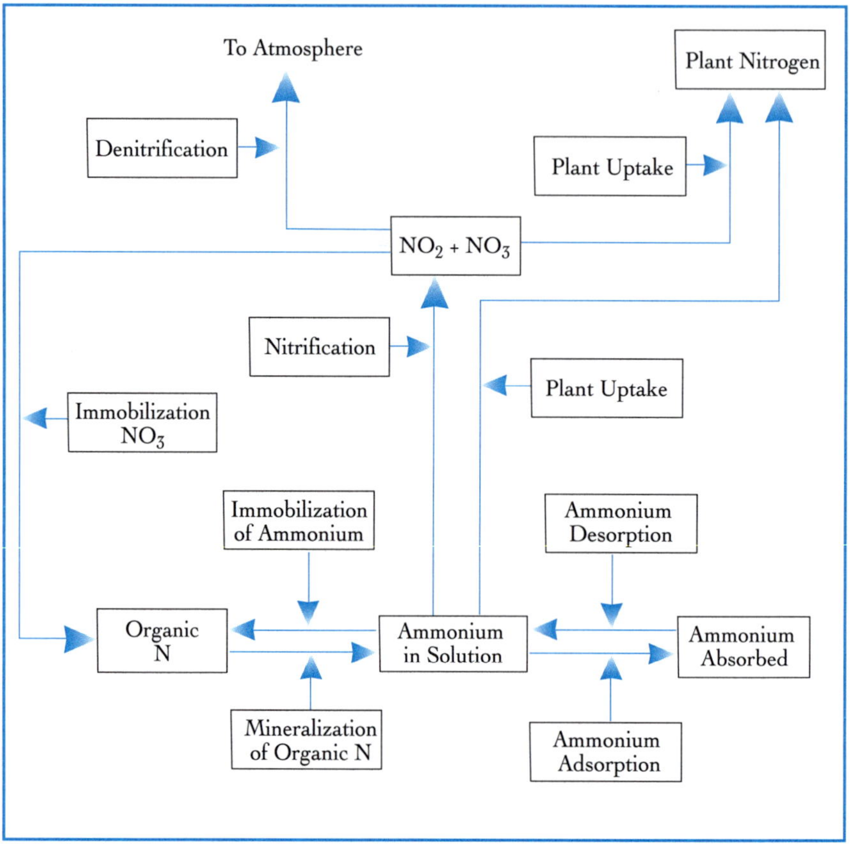

Figure HSP 3 Flow Diagram for Phosphorus Reactions

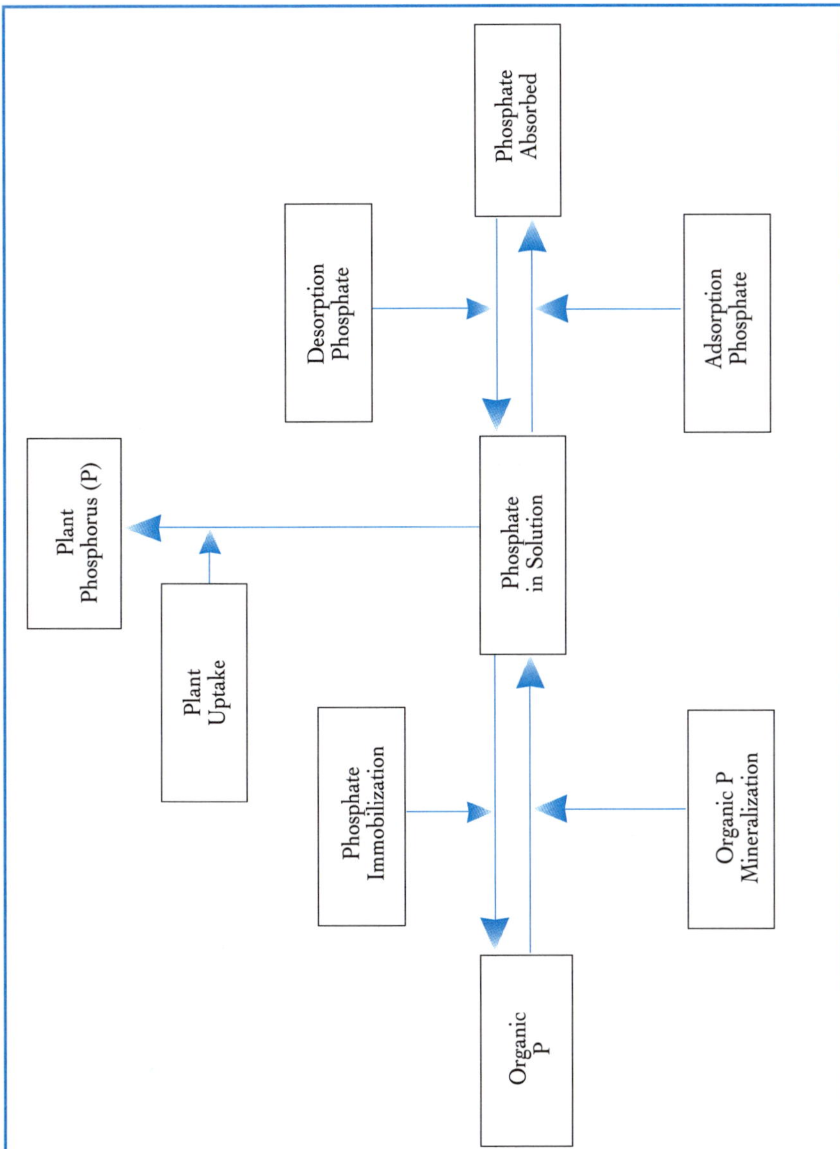

Figure HSP 4 Flow Diagram for Solids

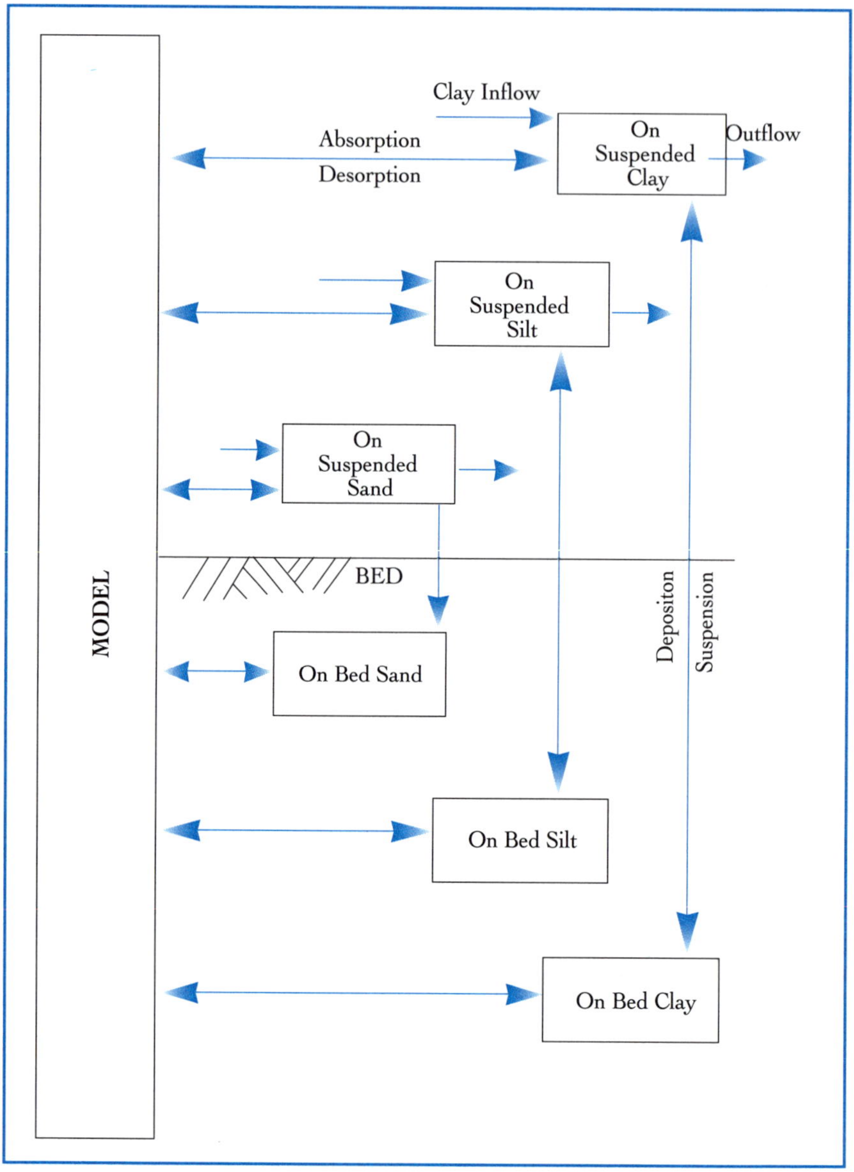

MIKE SYSTEM

Source: Danish Hydraulic Institute
Agern Alle 5
DK – 2970, Horsholm
Denmark
Proprietary model. Technical Reference Manuals are available.

Model Hardware Requirements: The minimum hardware requirements are a 586 computer with mathematical processor, 32k RAM, 5 Gbyte hard disk, and printer and plotter.

Professional Expertise Requirements: Professional with experience in numerical modeling, hydrodynamics, and water quality.

Receiving Water: MIKE SYSTEM consists of different models (MIKE 11, MIKE 21, MIKE 3, MIKE SHE, MOUSE and MIKE BASIN); consequently, the models can be used for rivers, lakes, estuaries, and coastal regions.

Spatial Characteristics: One, two, or three dimensions. MIKE 11 is a one-dimensional modeling system, MIKE 21 a two-dimensional modeling system, MIKE 3 a three-dimensional modeling system, MIKE SHE an integrated hydrological modeling system, MOUSE an integrated modeling package for urban drainage and sewer system, and MIKE BASIN an integrated water resources management and planning system.

Temporal Characteristics: Dynamic and time-variable.

Water Quality Parameters Simulated: Water level, velocity, temperature, dissolved oxygen, biochemical oxygen demand, nitrate, ammonia, dissolved conservative substances (e.g., salinity), and dissolved substances with a decay term (e.g., indicator bacteria).

Model Applications

MIKE 11 NAM (the hydrological module of MIKE 11 system): Data inputs required are catchment topography, precipitation, potential evapo-transpiration, and temperature. Model input parameters required are overland flow coefficient, interflow time constant, threshold value overland flow, threshold value interflow, time constants, specific yield of groundwater reservoir, time constant baseflow, storage including snow, surface, lower zone, and groundwater. Model outputs include daily runoff, variation of soil moisture content, and ground water recharge.

MIKE 11 HD (the hydrodynamic module of MIKE 11 system): The hydrodynanic model is for rivers based on the numerical solution of the coupled partial differential equations for continuity, momentum, and dissolved substance dispersion and transport. Data inputs required are river divided into reached, river cross-sectional data, upstream boundary conditions, and tributary and other discharges and withdrawals. Model input parameters required are time step and duration of simulation, bottom roughness coefficients for channel and flood plains, initial conditions, dispersion coefficients, and first-order decay rate. Model outputs include river water level, river flow, and accumulated flow and concentrations of dissolved substances, either conservative or affected by first-order decay only.

MIKE 11 WQ (the water quality module of MIKE 11 system): Data input required are river flow and water level from the hydrodynamic model, loadings from upstream boundary, tributaries and discharges, and initial conditions in the river. Model input parameters required are degradation constants, sedimentation and resuspension rates, photosynthesis and respiration rates, critical velocities for sedimentation, sediment oxygen demand, re-aeration rates, oxygen yield and uptake rates for ammonia and phosphorus, reaction rates for nitrification and denitrification, and adsorption/desorption of phosphorus on suspended solids. Model outputs include dissolved oxygen, biochemical oxygen demand, ammonia, nitrate, phosphorus, and temperature at river cross-sections.

MIKE 21 HD (the dynamic two-dimensional hydrodynamic module of MIKE 21 system): The model can be used for rivers, lakes, estuaries, and coastal regions based on the numerical solution of the coupled partial differential equations for continuity, momentum, and dispersion and transport of dissolved substance. Data inputs required are similar to MIKE 11 above but in two dimensions plus some other inputs like the velocity (flux) distribution and bathymetry. Model input parameters required are similar to MIKE 11. Model outputs include water levels, depth-averaged velocity vectors, flow, and accumulated flow.

MIKE 21 WQ (the dynamic water quality module of MIKE 21 system): Data inputs required include the output from the hydrodynamic model. MIKE 21 WQ Model inputs parameters required are similar to MIKE 11 but in two dimensions. Model outputs are similar to MIKE 11 but in two dimensions.

MIKE 3 PA (one of the sediment/particle transport module of MIKE 3 system): The model can simulate the transport and fate of dissolved and suspended substances discharged or accidentally spilled into rivers, lakes, estuaries, coastal areas, or open seas. The transport can be in two or three dimensions. This is a Lagrangian model for the advection and dispersion resulting from random processes of released uniformly distributed particles. It requires current velocities and water levels in time and space for the computational grid. The model can predict any conservative and decaying substances. Model inputs include bathymetry, bed friction coefficients, source data, wind data, dispersion coefficients, velocity depth profiles, and decay coefficients. The following discussions do not relate to a Lagrangrian model.

Note: the following three models are not DHI MIKE system family.

TIDEFLOW-3D: Dynamically simulates currents, dissolved solids, and temperature.

XXFLOW-3D: Simulates advection and transport of mud and single pollutants and interacting water quality parameters using the output of TIDEFLOW-3D.

XXPLUM-3D: Random walk models for sediment and dissolved substance plumes.

Model Attributes

The area must be divided into catchment areas. Topographical maps and other information related to runoff must be obtained for the catchment areas. The rivers are divided into branches of similar hydraulic characteristics (alignment, river cross-sections, slope, and river bottoms). In the case of the two-dimensional model, the area is divided into grids. Model predictions are for the branch boundaries or grid intersections. The partial differential equations are solved numerically using an implicit finite difference scheme on the computational grid. The point and non-point discharges must be quantified and properly located in the river reach or grid. Similarly, upstream boundary and initial conditions must be defined as well as the initial water quality concentrations.

Physical Aspects of the Model

All predicted parameters are dynamic or time-varying. The user can select among six levels of water quality descriptions, from the simple dissolved oxygen and biochemical oxygen demand to a model that includes the most important dissolved oxygen processes. An example of the model setup is shown in Figure 4.13.

Dissolved Oxygen

The various dissolved oxygen processes included in the two-dimensional model are shown in Figure MIK 1. The processes shown in Figure MIK 1 are superimposed on the simulations from the hydrodynamical model.

Nutrients – Nitrogen and Phosphorus

The nitrogen processes in the model are shown in Figure MIK 2. The phosphorus processes in the model consist mainly of phosphorus adsorption and desorption from suspended sediments. These processes are superimposed on the simulations from the hydrodynamical model.

Two-Dimensional Model

The velocity vectors are simulated in two dimensions. Figure 4.15 is an example of the simulation of the maximum ammonia concentration.

Error and Sensitivity Analysis

There are no recommendations or discussion for determining the errors and sensitivity of the model input parameters. The sensitivity of the model simulations to various input parameters is useful to the model user for the model calibration process. The model user can use recognized statistical methods to conduct a tailored sensitivity analysis and determine the sensitivities.

Calibration

Model calibration is discussed in the Chongqing report. A calibration data set is selected and the model parameters adjusted to obtain a match between measured and simulated values. This process is carried out for water levels, flows, accumulated flows, temperature, and the other simulated water quality parameters. The parameters that can be adjusted in the calibration process are identified in the model documentation.

Verification

Model verification is discussed in the Chongqing report. In this process, the calibrated model simulated values are compared with another data set. Any significant differences are investigated. If the deviations are unacceptable large, it may be necessary to recalibrate the models.

General Comments

The model in its two-dimensional form is very complex, requiring many different data inputs. The complexity of the model was appropriate for the Chongqing project, where the receiving water consists of two major rivers with two tributaries upstream of the Three Gorges Dam. These rivers have 50 major domestic wastewater outfalls, 12 major industrial outfalls, and 15 non-point source loadings in the Chongqing region. There are also large loadings to the Yangtze River upstream of the Chongqing area. However, complex models are more difficult to calibrate and verify because of the number of input parameters and data that must be provided to the model. In the case of Chongqing, many of the simulated values were within 5 percent, and some differed by 200 to 300 percent. It is important that the model user know in what instances the simulated and measured values have significant differences and interpret the simulations in an appropriate manner. If the model is correctly formulated, it can be argued that it must be possible to calibrate the model properly if the accurate calibration data are available. Usually there are insufficient data for the calibration process and, even if these data are available, the calibration process may still not be perfect because of the number of different combinations of input parameters that must be assessed and analyzed. Time and cost usually limit the calibration process.

Figure MIK 1 Dissolved Oxygen Processes

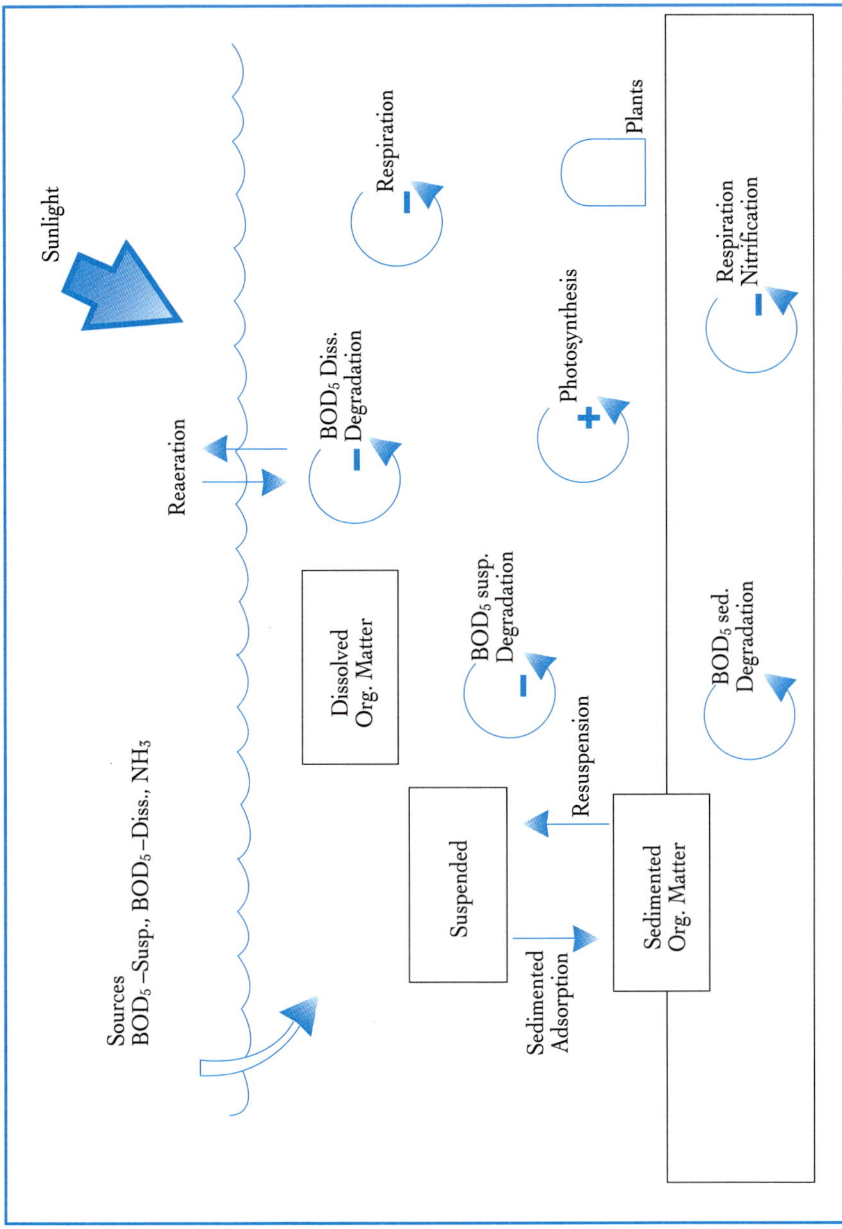

Figure MIK 2 Nitrogen Processes

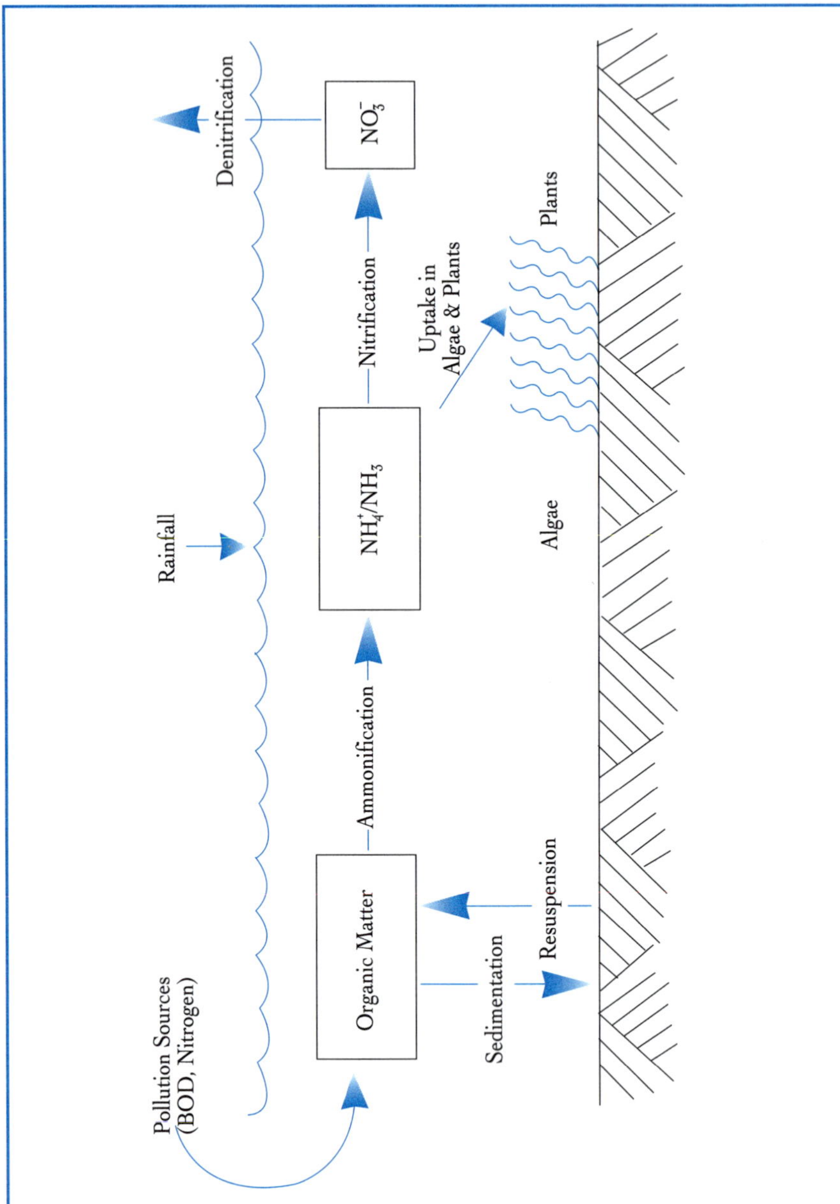

QUAL2E & QUAL2E-UNCAS (6 APRIL 1999)

Source: U.S. Environmental Protection Agency
Office of Research and Development
Environmental Research Laboratory
960 College Station Road
Athens, Georgia 30605-2700
Send two blank 3.5" HD diskettes. No other costs.
Manual available from NTIS.
Tel: 703-605-6000; ask for EPA/600/3-87/007, May 1987.

Model Hardware Requirements: PC with at least 256K memory with input disk file and line printer or disk for output.

Professional Expertise Requirements: A university degree with courses in hydrodynamics and environmental engineering.

Receiving Water: Primarily narrow rivers with tributaries.

Spatial Characteristics: One-dimensional – downstream direction.

Temporal Characteristics: Steady state for one solution – does not have time as a variable; however, model can be run repeatedly to simulate time variability.

Water Quality Parameters Simulated: Dissolved oxygen, biochemical oxygen demand, nutrients, indicator bacteria, conservatives, and non-conservatives.
Uncertainty analysis is available for model parameters in the form of sensitivity analysis, error analysis, or Monte Carlo simulation.

Model Applications: To determine the impact of waste loadings from both point and non-point sources on the receiving water quality.

Model Attributes

The river is divided into reaches that have similar hydraulic characteristics (maximum 25 reaches). Each reach is divided into computational elements of equal length. Elements at the headwaters, at junctions, at inputs or withdrawals, and the last element are identified in the model (Figures QUA1 and QUA 2). The one-dimensional advection-dispersion mass transport equation is numerically integrated over space and time for each water quality constituent. Equations are solved by finite difference methods.

Physical Aspects of the Model

The river flow is assumed constant at the headwater element. The mass transport equation requires that stream cross-sectional area, mean stream velocity, downstream dispersion coefficient, and the volume of water in the element be known. The mean velocity, cross-sectional area, and depth in the reach are determined using the power functions of the flow or particular cross-sectional data (e.g., trapezoidal, rectangular, etc.). Three different options can be selected for determining the downstream dispersion, and some of these are dependent on the channel roughness. A table of channel roughness is provided in the manual, and some typical dispersion coefficients in rivers are provided in a table. The rivers are U.S. rivers, but details on the river depth, width, and mean velocity are provided in the table so that these data can be considered universal.

Primary productivity

Algal bio-mass is determined as a function of the chlorophyll a measurement. Productivity is determined from light functions (three options) or light averaging functions (four options) and algal self-shading.

Algal-limiting nutrient relationships are developed using Michaelis-Menten ratios.

Nitrogen cycle

The equations for organic nitrogen, ammonia nitrogen, nitrite nitrogen, and nitrate nitrogen are developed and solved. These relationships require algal bio-mass.

Phosphorus cycle

The equations for organic phosphorus and dissolved phosphorus are developed and solved. Algal bio-mass is an important component of the equation.

Carbonaceous BOD

The equations for BOD assume a first-order reaction to describe the de-oxygenation of BOD. This consists of a rate reduction, which is temperature-dependent, and settling of the BOD particles.

Dissolved oxygen

The rate of change of dissolved oxygen depends on re-aeration, photosynthesis, oxygen of the incoming flow, oxidation of carboneous and nitrogenous organic matter, and benthic oxygen demand and respiration.

Eight different methods exist for determining re-aeration as well as a special re-aeration method for dams. Default values for all of the coefficients are presented in a table in the manual. Figure 2.1 shows the dissolved oxygen processes included in the model.

Indicator bacteria

Bacterial densities are simply determined using a first-order decay superimposed on the advection and dispersion predictions.

Temperature

Temperature is determined by a heat budget which considers solar radiation (short- and long-wave and reflection), convective heat, evaporation losses, and heat flux. This component of the model has been found to be very effective.

Tabular values of the default options for all the coefficients in the model are provided, as well as a range of measured values.

The manual contains good schematics that are helpful for visualizing how the model functions. Details on the finite difference method used to solve the equations are well presented in the manual.

Model Error and Sensitivity Methods

The model error and sensitivity methods are presented as an option in the model application. The three types of error analysis are as follows:

Sensitivity analysis: In the model, the sensitivity of the model predictions can be assessed single inputs or groups of inputs or factorial design.

First-order error analysis: It is possible to obtain a table of sensitivity coefficients (normally distributed) as a percentage of change in the output variance. It is also possible to obtain the components of variance of each output variance as the percentage of the output attributed to each input variable.

Monte Carlo simulation: It is possible to run the model in a Monte Carlo simulation mode.

An excellent table is provided on the model uncertainty associated with the variance of the input variables. As a typical percentage, these can be summarized as follows:

Temperature	2–3%
BOD	0–20%
Dissolved oxygen	5–10%
Nitrogen	15–30%
Phosphorus	15–40%
Algae	10–25%
Bacteria	0.2–0.35 log
Conservative	5–10%
Cohesive sediments and flocs	50–100%

The user manual is easy to use and is presented in a step-by-step manner.

General Comments

QUAL2 is one of the most extensive water quality prediction models used. The configuration of the model is flexible, so that the user can change it for special applications. This model is so popular that it has been incorporated in many other models in one form or another. The model can be used with very little site-specific data, which is its advantage and disadvantage, because the model can be used by personnel with no expertise in receiving water quality, hydraulics, or modeling/prediction.

Figure QUA 1 Stream Network of Computational Elements and Reaches

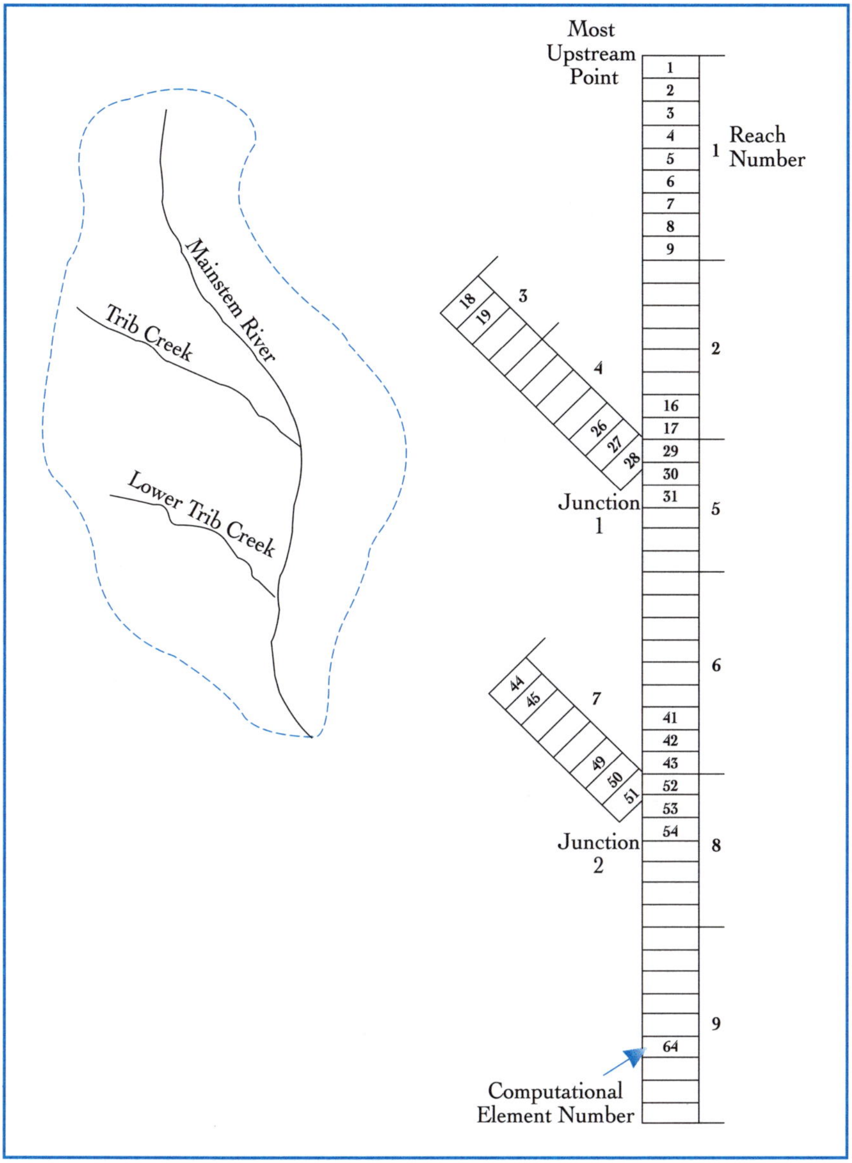

Figure QUA 2 Discretized Stream System

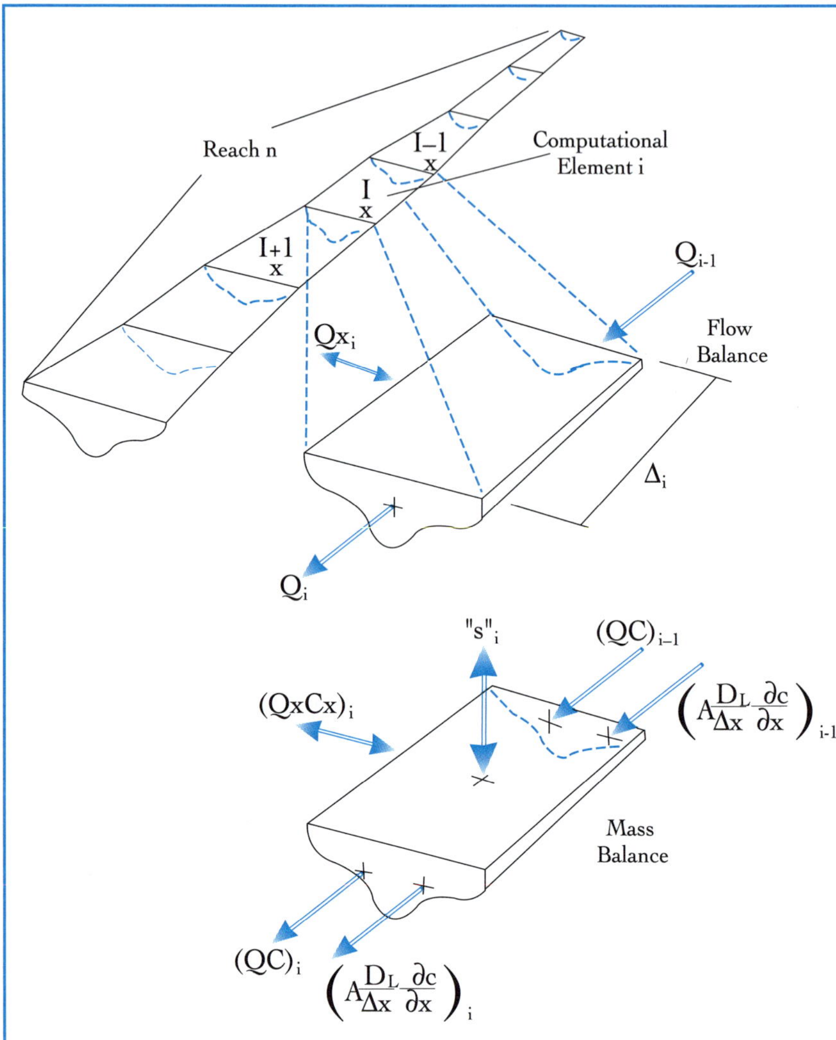

STORM WATER MANAGEMENT MODEL (SWMM) VERSION 4 PART A: USER'S MANUAL

Source: U.S. Department of Commerce
National Technical Information Service
5285 Port Royal Road
Springfield, VA 221161
Tel: 703-605-6050
Reference: EPA/600/3-88/001a, June 1988
or
Environmental Research Laboratory
U.S. Environmental Protection Agency
960 College St. Road
Athens, Georgia 30605.
Paper copy of the user manual is available from the addresses above. The program is free of charge from the Athens address above, but the request must be accompanied by two HD 3.5" disks.

Model Hardware Requirements: RAM 512 Kbytes, diskette, hard drive of 3Meg or more, numeric co-processor, DOS, and FORTRAN 77.

Professional Expertise Requirements: User should have a science degree, be knowledgeable in modeling techniques, and have an appreciation of modeling physical processes. SWMM is a large, relatively sophisticated hydrologic, hydraulic, and water quality simulation program.

Spatial Characteristics: One dimension.

Temporal Characteristics: Dynamic – time is a variable.

Water Quality Parameters Simulated: None. SWMM is set up to link with WASP4 and DYNHYD.

Model Applications: Predicts runoff for complicated urban areas with sewer systems from rainfall and topographical information in the catchment.

Model Attributes

This is a very complex model that predicts runoff in real time from rainfall data for an urban area. The model has been extensively used for urban areas and is one of the best models for predicting runoff from urban areas with a well-developed sewer system. The model can be used as a black box runoff prediction model with minimal input data if calibration and verification data are available. Figure SWM 1 shows the components of the model; Figure SWM 2 shows a typical setup for a model input.

Model Error and Sensitivity

The quantity prediction from SWMM may be very accurate, with little calibration because of the model structure. The error can be determined by using verification data.

General Comments

The model is very useful for generating flows in pipe networks in urban areas. The model can handle any hydraulic cross-section as well as different flow regimes (Figure SWM 3). The model has good hydraulic and sediment dynamics sub-routines for open channel flow and good backwater computations in the EXTRAN Addendum (EPA/600/3-88/001b). The model has a dry weather flow module, which is very useful for water quality management because it is the dry weather flow periods that frequently stress the receiving water quality. The manual is easy to use, and the model processes are described well.

Figure SWM 1 Relationship Among SWWM Blocks

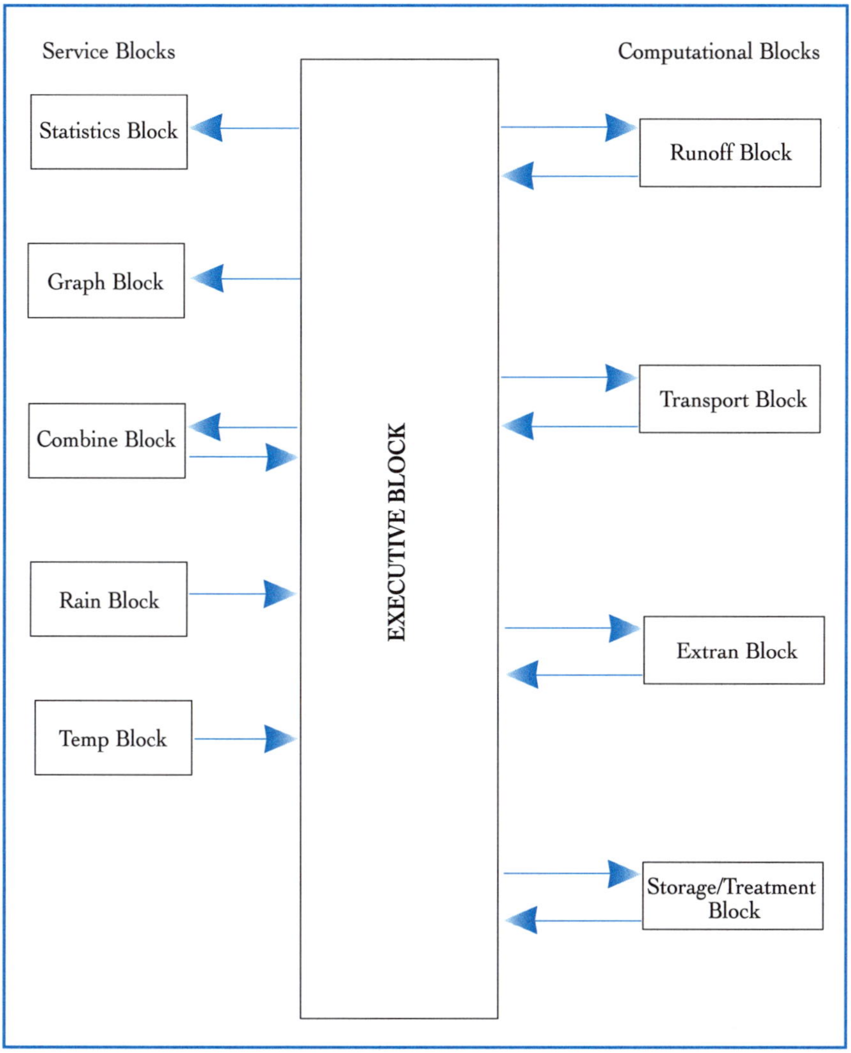

Figure SWM 2 Northwood (Baltimore) Drainage Basin "Coarse" Plan

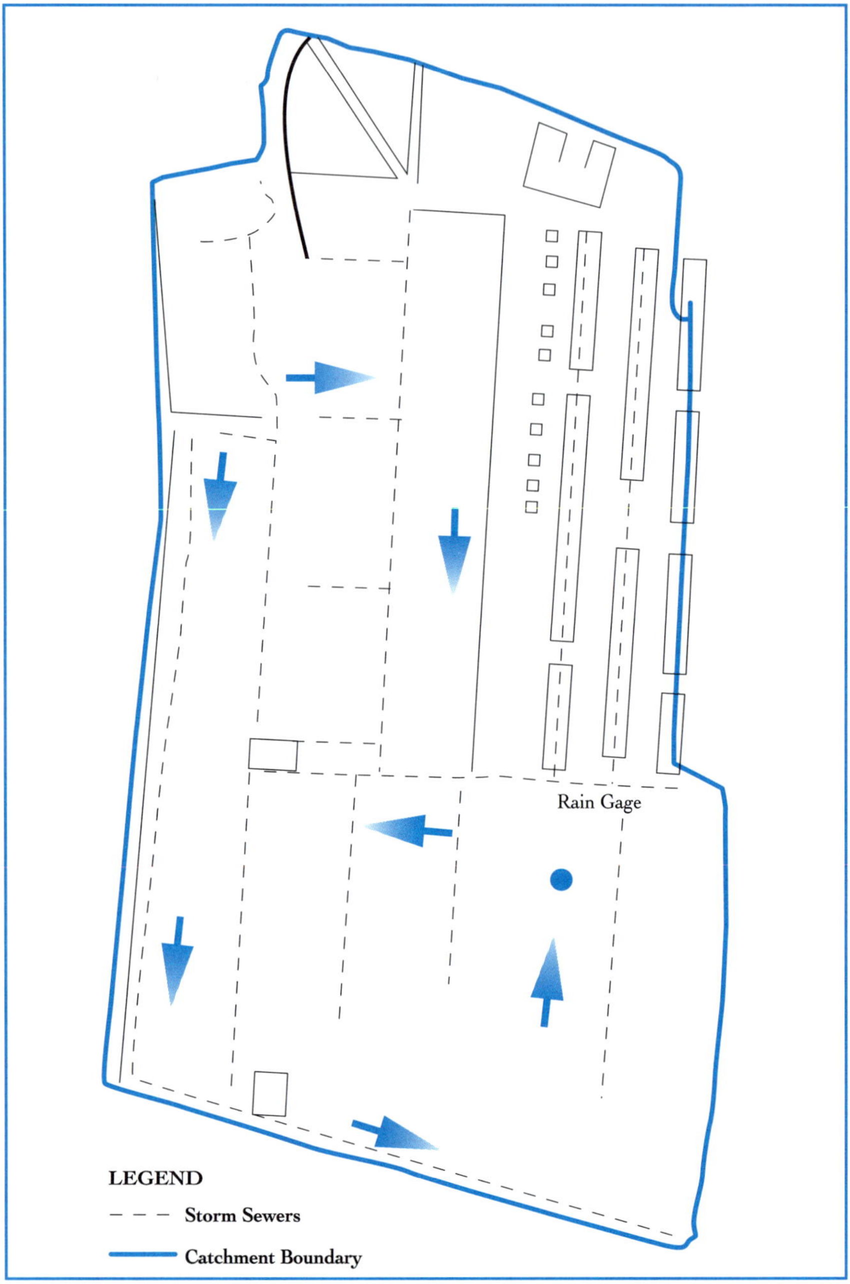

Figure SWM 3 Special Hydraulic Cases in EXTRAN Flow C

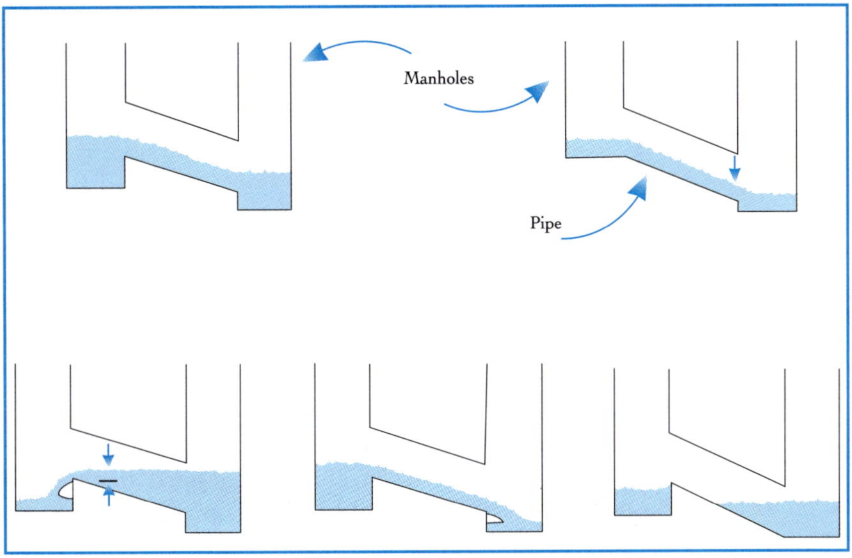

TRISULA - DELWAQ
DELFT HYDRAULICS

Source: Delft Hydraulics
Rotterdamseweg 185
P.O. box 177
2600 MH Delft
The Netherlands
Tel: 31 15 2858585
Email: info@widelft.ni
Details on the models are available at the address above.

Model Hardware Requirements: Unknown because it is a proprietary model. Presumably, a recent personal computer with 32K RAM and 2Meg hard drive with math co-processor would be appropriate.

Professional Expertise Requirements: User should have a science degree, be knowledgeable in modeling techniques, and have an appreciation of modeling physical processes.

Spatial Characteristics: Two and three dimensions.

Temporal Characteristics: Dynamic.

Water Quality Parameters Simulated: Inorganic and organic sediment kinetics, nutrient kinetics, phytoplankton, nitrification and denitrification, mineralization, oxygen and biochemical oxygen demand, heavy metals, indicator bacteria, and oils.

Model Applications: Lakes, estuaries, and coastal regions.

Model Attributes

This model consists of equations that are solved numerically using finite element techniques. Figure TRI 1 shows the model modules and the input and output data. TRISULA is the hydrodynamic model and DELWAQ the water quality model. Figure TRI 2 shows the water quality substances and processes included in DELWAQ. Figure TRI 2 clearly shows that the processes simulated are comprehensive and include surface processes like re-aeration, denitrification, and volatilization as well as the bottom sediment processes like sedimentation, mineralization, resuspension, and burial. The water processes include all the major components of the primary productivity system as well as dissolved and detritus of the nutrients, spill, and first-order decay processes. Details on the operation and application of the models are available from the owners.

Model Error and Sensitivity

The total concentration levels are calibrated as well as the water quality and ecological processes. The various model coefficients and rate constants are systematically adjusted in the calibration process. Details on the calibration process are discussed in the project report for Hangzhou Bay. Figure 4.12 shows an example of the calibration process. In the figure, the differences between the simulated and measured salinities in the bay are presented. The calibration process, coupled with some sensitivity analysis for the user-provided coefficients and rate constants, should quantify the precision of the model simulation for any application.

General Comments

The models have been extensively used on large water resource projects throughout the world. Further details on the models can be obtained from Delft Hydraulics.

Figure TRI 1 General Structure of the Modeling Framework

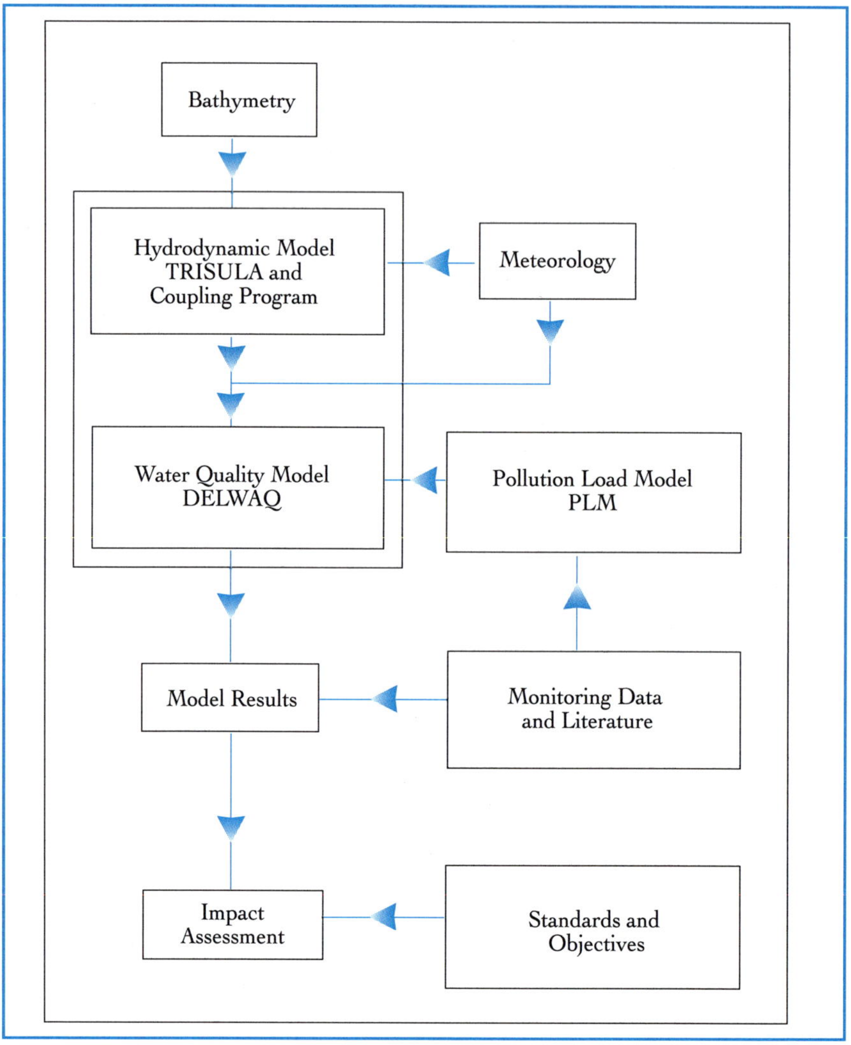

Figure TRI 2 Model Processes

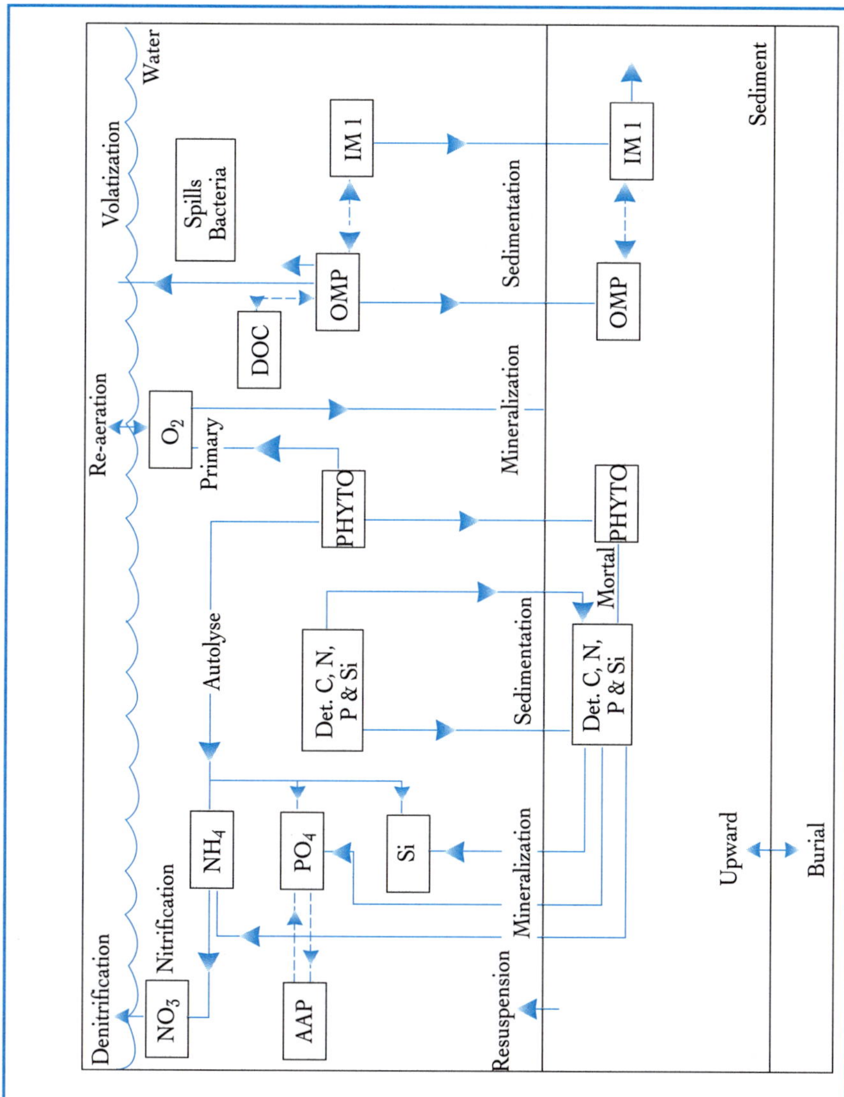

WQRRS
Water Quality for River-Reservoir Systems

Source: Hydrologic Engineering
U.S. Army Corps of Engineers
609 Second Street
Davis, CA 95616
Tel: 916-756-1104
Program and manual are available on the internet, password 9167561104. Contact the above for other information.

Model Hardware Requirements: A recent Pentium personal computer with 32K RAM and 2Meg hard drive with math co-processor with input disk file and line printer or disk for output.

Professional Expertise Requirements: A university science degree with experience in hydrodynamical modeling and water quality modeling.

Receiving Water: River-reservoir systems.

Spatial Characteristics: One and two dimensions.

Temporal Characteristics: Dynamic – time is a variable.

Water Quality Parameters Simulated: Dissolved oxygen, biochemical oxygen demand, nutrients, algal bio-mass, temperature, indicator bacteria, conservatives and non-conservatives, reservoir nutrients and algal productivity, as well as the interaction of reservoir flow and temperature.

Model Applications: Reservoir management for releases times and volume rates to maintain water quality in the reservoir.

Model Attributes

This model is an extension of the CE-QUAL-W2 reservoir model that predicts the input quality and quantity for the river system tributary to the reservoir. The river portion of the model is similar to QUAL2. WQRRS integrates the outputs from the river model into the reservoir model.

Physical Aspects of the Model

The physical characteristics of the model are similar to those of QUAL2 and CE-QUAL-W2, and the outputs from the river model are integrated in the reservoir model.

Model Error and Sensitivity Methods

The model error and precision must be determined by the model user.

General Comments

This model is based on the QUAL2 and CE-QUAL-W2 models, which have both been extensively used in the Western hemisphere.

Glossary

Advection: The concentration multiplied by the velocity vector in the mass balance equations.

Bathymetry: Measurements of the water depths.

Biochemical oxygen demand (BOD): The concentration of dissolved oxygen required to oxidize organic and inorganic substances expressed at a water temperature.

Buoyancy: A difference in density between two fluids.

Calibration: The process of adjusting the coefficients and parameters in a water quality model so that the predictions correspond to a set of measurements.

cfu: Colony-forming units.

Coefficient: A dimensionless constant number.

Cohesive: Inorganic particles less than 2um in size that have large surface areas.

Concentration: The mass of a substance in the mass of water.

Denitrification: The reduction of nitrate and nitrite to nitrogen with the release of oxygen. This process is achieved with bacteria and enzymes.

Density stratification: Layers of different density in the fluid.

Detritus: A mixture of organic and inorganic particles suspended in a fluid.

Differencing: Subtracting the predictions from two separate model applications.

Diffuser: A device for spreading the effluent discharge in the receiving water, typically using numerous smaller diameter ports spaced at various distances along the outfall pipe (multiport diffuser).

Dilution: The mixing of a volume of water with a concentration of a substance with another volume of water at a lower concentration.

Dispersion: The spreading of a dissolved substance in water.

Dissolved oxygen (DO): The concentration of dissolved oxygen in the liquid expressed in mg/L.

Dynamic: Varies with time.

E. coli: A bacterial species that exists in great numbers in the intestines; consequently, it can be used to quantify the presence of fecal matter in the receiving water.

Eulerian: A fixed reference system, normally three axes at right angles to each other with a fixed location.

Eutrophication: Providing an abundance of needed nutrients to aquatic plants.

Fecal coliform: A group of intestinal bacteria that can be used to quantify the presence of fecal matter in the receiving water.

Geometric mean: The antilog of the mean of the logs.

Herbicide: A chemical substance that destroys living plants.

Hydrodynamic: Relating to water movement.

Hydrological: Relating to the volume and flow of water on or below the ground originating from precipitation events.

Initial dilution: The dilution achieved by the kinetic energy in the effluent discharge.

LCs: The concentration that causes 50 percent fish mortality in 96 hours.

Lagrangian: A moving reference system, normally three orthogonal axes at right angles to each other with an origin that moves as a function of time.

Loading: The concentration of a water quality parameter multiplied by the wastewater flow in mass loadings per unit of time.

Macrophyte: Rooted aquatic plant.

Nitrification: The oxidation of ammonia to nitrate. The process is achieved by bacteria.

Nutrients: Chemicals necessary for the growth of aquatic plants and bacteria. (Only phosphorus and nitrogen are considered in this guide.)

PAH: Polycyclic aromatic hydrocarbons.

Pesticide: A chemical/biological substance that destroys insects.

pH: The acidity.

Photosynthesis: Aquatic plant process of using sunlight energy measured as the production of oxygen.

Physical shear: The gradients of velocity in the fluid.

Phytoplankton: Microscopic free-floating aquatic plants.

QA/QC: Quality assurance (generally the accuracy) and quality control (generally the precision).

Replicate samples: Two separate samples collected at the same location, same depth, and at the same time.

Respiration: Aquatic plant process in the absence of sunlight measured as the production of carbon dioxide.

Secchi disc: A metal disc that is painted in alternating white and black quadrants.

SOD: Sediment oxygen demand

Split samples: Dividing a single sample into two or three parts. Each part is called a split.

Steady-state: Does not vary with time.

Stochastic: Random or probability.

Streamlines: A line that is parallel to the local velocity vector.

Surface re-aeration: The increase of the dissolved oxygen concentration through atmospheric processes, both physical and chemical.

Suspended solids: The mass of organic and inorganic particles suspended in water.

T90: The time required for a 90 percent reduction in concentration.

Total coliforms: A group of intestinal bacteria that can be used to quantify the presence of fecal matter in the receiving water.

Transport: The movement of a parcel of water.

Verification: The process of comparing the predictions of a calibrated water quality model with a set of measurements that were not used in the calibration process.

Volatilization: The process of a change of state in fluids between liquid and a gas.

Zooplankton: Microscopic aquatic animals.

References

Ackers, P. and W.R. White, 1973. Sediment transport: new approach and analysis. *American Society of Civil Engineers, Hydraulics Division* 99(HY11), 2041-2060.

Alsaffar, A.M., 1966. Lateral diffusion in a tidal estuary. *Journal of Geophysical Research* 71(24): 5837-5871.

Barnwell, T.O. and P.A. Krenkel, 1982. The use of water quality models in management decision making. *Water Science Technology* 14: 1095-1107.

Benarie, M.M., 1987. Editorial: The limits of air pollution modeling. *Atmospheric Environment* 21: 1-5.

CORMIX, 1993. Expert system for hydrodynamic mixing zone analysis for multiport discharges. Environmental Research Laboratory, U.S. Environmental Protection Agency, Athens, Georgia 30605. EPA/600/3-93/007.

CORMIX1, 1990. Expert system for hydrodynamic mixing zone analysis of conventional and toxic submerged single port discharges. Environmental Research Laboratory, U.S. Environmental Protection Agency, Athens, Georgia 30605. EPA/600/3-90/012.

Denman, K.L., A. Okubo, and T. Platt, 1977. The chlorophyll fluctuation spectrum in the sea. *Limnology and Oceanography* 22: 1033-1038.

Dewey, R.J., 1984. Application of stochastic dissolved oxygen model. *Journal of Environmental Engineering* 110(2): 412-429.

Effler, S.W., 1985. Attenuation versus transparency. *Journal of Environmental Engineering* 111(4): 448-459.

Ellis, H.M., et al., 1980. Comparison of predicted and measured concentrations for 58 alternative models of plume transport in complex terrain. *Journal of Air Pollution Control Association.* June.

Gannon, J.J. and M.K. Busse, 1989. *E. coli* and enterococci in urban stormwater, river water and chlorinated treatment plant effluent. *Journal of Water Research* 7: 789-798.

Gowda, T.P.H. and J.D. Lock, 1985. Volatilization rates of organic chemicals of public health concern. *Journal of Environmental Engineering* 111(6); 755-776.

Harris, G.P., 1980. Temporal and spatial scales in phytoplankton ecology, mechanisms, mehtods, models and management. *Canadian Journal Fisheries Aquatic Sciences* 37: 877-900.

Huang, J.C. and F.C. Monastero, 1982. Review of the state-of-the-art of oil spill simulation models. American Petroleum Institute, Washington, D.C.

Lewis, W.K. and W.G. Whitman, 1924. Principles of gas absorption. *Industrial and Engineering Chemistry* 16(12); 1215-1220.

Metcalf & Eddy, 1991. Wastewater Engineering. McGraw-Hill Inc., New York.

Muellenhoff, W.P., A.M. Solate, Jr., D.J. Baumgartner, M.D. Schuldt, L.R. Davis, and W.E. Frick, 1985. Initial mixing characteristics of municipal ocean discharges. Volume 1 – procedures and applications. Environmental Research Laboratory, U.S. Environmental Protection Agency, Narragansett, RI 02882. EPA/600/3-85/073a.

National Council of the Paper Industry for Air and Stream Improvement Inc., 1982. A review of ultimate BOD and its kinetic formulation for pulp and paper mill effluents. Technical Bulletin 382.

National Council of the Paper Industry for Air and Stream Improvement Inc., 1986. A review of the separation of carbonaceous and nitrogenous BOD in long-term BOD measurements. Technical Bulletin 461.

NHW, 1992. Guidelines for Canadian Recreational Water Quality. Canadian National Health and Welfare. Canadian Government Publishing Centre, Supply and Services Canada, K1A 0S9.

Okubo, A., 1971. Oceanic diffusion diagrams. *Deep Ocean Research* 18: 789-802.

Palmer, M.D., 1981. Some measurements of near surface turbulence in the depth direction and some phytoplankton distribution implications. *Journal Great Lakes Research* 7(2): 171-181.

Palmer, M.D., 1984. Surface streaking and fecal coliform levels. *Canadian Journal of Civil Engineering* 11(2): 351-355.

Palmer, M.D., 2000. Analyses of sediment bacteria monitoring data from two deep ocean raw wastewater outfalls, Victoria, BC. *Canadian Water Resources Journal* 25(1): 1-18.

Palmer, M.D. and L. Burrelle, 1996. Aerated stabilization basin floc size treatment implications. *Water Quality Research Journal Canada* 31(4): 787-799.

Palmer, M.D. and R.J. Dewey, 1984. St. John's harbor water quality model. *Canadian Journal of Civil Engineering* 11: 65-73.

QUAL2E, 1987. The enhanced stream water quality models QUAL2E and QUAL2E-UNCAS: documentation and user manual. Environmental Research Laboratory, U.S. Environmental Protection Agency, Athens, Georgia 30605. EPA/600/3-87/007.

Rasmussen, H. and H.M. Badr, 1979. Validation of numerical models of the unsteady flow in lakes. *Applied Mathematical Modelling* 3: 416-420.

Sharp, J.J. and E. Moore, 1987. Estimation of dilution in buoyant effluents discharged into a current. *Proceedings Institution Civil Engineers*, Part 2, 83: 181-196.

Sharp, J.J. and E. Moore, 1989. Marine outfall design-computer models for initial dilution in a current. *Proceedings Institution Civil Engineers*, Part 1, 86: 953-961.

Simons, T.J., 1985. Reliability of circulation models. *Journal Physical Oceanography* 15(9): 1191-1204.

SWMM, 1988. Storm water management model, version 4. Part A: User's manual. Environmental Research Laboratory, U.S. Environmental Protection AgencyAthens, Georgia 30605. EPA/600/3-88/001a.

Springthorpe, V.S., Chi Leong Loh, W.J. Robertson, and S.A. Sattar, 1993. In situ survivial of indicator bacteria, MS-2 phage and human pathogenic viruses in river water. *Water Science Technology* 27(3-4): 413-420.

HSPF, 1983. Hydrological simulation program – FORTRAN: Users' manual for release 8.0. Environmental Research Laboratory, U.S. Environmental Protection Agency, Athens, Georgia 30605. EPA/600/3-84-066.

Van Der Kooij, L.A., D. Van De Meent, C.J. Van Leeuwen, and W.A. Bruggeman, 1991. Deriving quality criteria for water and sediment from the results of aquatic toxicity tests and product standards: application of the equilibrium partitioning method. *Water Research* 25(6); 697-705.

Venkatesh, S., 1990. Model simulations of the drift and spread of the Exxon Valdez oil spill. *Atmosphere-Ocean* 28(1): 90-105.

WASP4, 1988. A hydrodynamic and water quality model – model theory, user's manual and programmer's guide. Environmental Research Laboratory, U.S. Environmental Protection Agency, Athens, Georgia 30605. EPA/600/3-87/039.

World Bank, 1998. Pollution Prevention and Abatement Handbook, 1998. World Bank, Washington, DC.

World Bank, 1991. Environmental Assessment Sourcebook. Volume II -Sectoral Guidelines. World Bank, Washington, DC.,Technical Paper Number 140.

World Health, 1988. Environmental Impact Assessment – Siting and Design of Submarine Outfalls. Monitoring and Assessment Research Centre, King's College London, University of London.

Yalin, M.S., 1977. Mechanics of Sediment Transport. Pergamon Press, New York.Ziclinski, P.A., 1988. Stochastic dissolved oxygen model. *Journal of Environmental Engineering* 114(1): 74-90.